I0220836

GROWING A FOOD FOREST

GROWING A FOOD FOREST

GROWING A FOOD FOREST

Trees, Shrubs, & Perennials That'll Feed Ya!

The Hungry Garden series #5

Rosefiend Cordell

GROWING A FOOD FOREST

Rosefiend Publishing.

Ordering information: For details, contact the publisher at hello@melindacordell.com

Cover design by Melinda R. Cordell

Evil Empire ISBN:

D2D ebook ISBN: 9781953196637

D2D ppb ISBN: 978-1-953196-77-4

First Edition: July 2023

10 9 8 7 6 5 4 3 2 1 blast off!

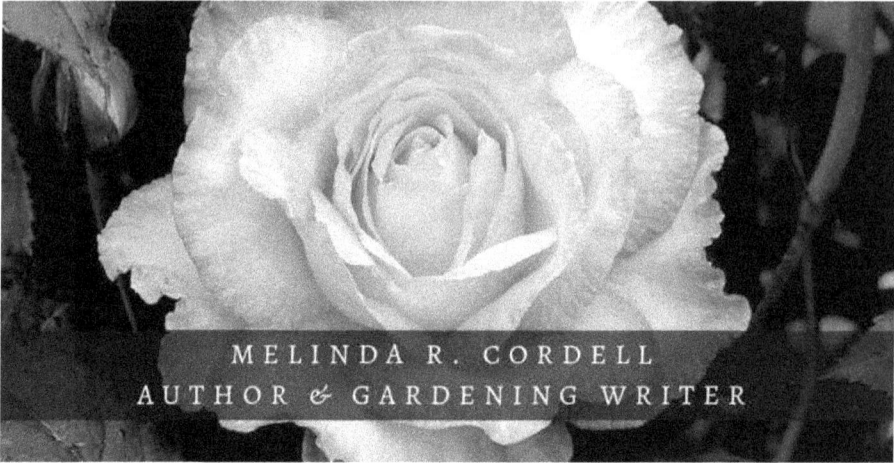

MELINDA R. CORDELL
AUTHOR & GARDENING WRITER

For more information (and books!), visit my website at
https://melindacordell.com/

Subscribe to my Newsletter
and get a free gardening book.

The Hungry Garden Series

FORTHCOMING BOOKS!

Survival Rations! – Foraging for Greens, Mushrooms, Berries, & Nuts
Book 6

Wildscaping – Using Native Food Plants to Create an Ecologically-Friendly Garden
Book 7

GROWING A FOOD FOREST

Table of Contents

INTRODUCTION

When I was a kid, I spent so much time in the forests around Nodaway, Missouri, where I grew up. I liked all wild spaces – I still do – and would spend many hours exploring the hills and floodplain. When I was older I would walk out to the Missouri River and watch it roll by. Sometimes the Missouri River would come to us, which is a whole 'nother story.

But the forests around my tiny town were my favorite places to explore. I loved looking for wildflowers. I could always find Dutchman's breeches growing alongside trout lilies. Mayapples and pawpaw trees also were congenial companions. I've never managed to find a ripe mayapple fruit – the animals always beat me to them – but I could find

1

ripe pawpaws in late fall, after the frost. I tried making pawpaw bread one time, using a recipe I found in the *Missouri Conservationist* magazine, but I didn't cook it long enough, so it was soggy in the middle.

There always seemed to be something to snack on, out in the woods. Mulberry trees dropped tons of purple fruits in June, or whitish-purple fruits if you found a white mulberry tree. Raspberries would show up in early July, much to my everlasting delight. Blackberries came later, but I avoided these because these brambles would lunge at your with knives. But in August, I could find dewberries growing along the ground – a nice, docile bramble with tasty berries, and I always wanted to find more.

Early in spring fern fronds would pop up with morel mushrooms – half the people in Missouri make a mad dash for the forests during morel season – then oyster mushrooms and other varieties. Wild strawberries spread everywhere, along with wild grapes in fall. Jerusalem artichokes – a type of sunflower with edible tubers – grew in the old cemetery at the top of the hill. Hickory and walnut trees grew up everywhere. Occasionally I'd find a persimmon tree, but I never could pick a properly ripe persimmon. They were always astringent and turned my mouth inside-out.

I thought that the Native Americans who used to live here had a pretty good deal, with all this food growing in the forest. I imagined that they had grown food in the forest at one time, using what was already available. They'd cut back the plants that were surrounding the food plants, and they could propagate and grow more food and medicinal plants to support the people who lived there or were moving through.

Now, years later, I realize that these crops had been planted, encouraged, and raised by the Native people who lived on this land for millennia, until many of them were pushed out of Andrew County (as well as the entire Platte Purchase) in 1840. Those who managed to stay eventually had their land stolen too, and were forced into Kansas, and from thence into different territories, far away from their old beloved homes. Hundreds of years of Native history were erased, and now the people who live here think *they're* the pioneers who settled an empty land, and there was nothing here before them.

When it comes to heirloom seeds and knowledge of how to grow crops in unique locations, Native folks understand the importance of all of this. The rest of us are trying to get caught up.

The forests tell a story of what had been – how things should be, really, especially while pollinators and other creatures are failing due to climate change and habitat loss. But wild areas can show us how things should be done.

In the end, permaculture done right is about creating an environment where plants can perpetrate themselves over time. Perhaps you have spent a lot of time hoeing and weeding your raspberry field, only to take a walk down a country road and find raspberries growing wild and bearing like crazy. Plants don't need as much work as you think if you give them the right conditions, and if you raise plants that do well in your specific climate.

Welcome to the wide wild world of food forestry!

P.S. If you're looking for more aesthetics and pretty stuff, I wrote a book called *Edible Landscaping: Foodscaping and Permaculture for Urban Gardeners* that covers that end of it a lot better.

Step 1: Grow a forest.
Step 2: ???
Step 3: Profit.

UNDERSTANDING AGROFORESTRY

The whole idea of creating forests that grow food goes under several different names – agroforestry ... silviculture ... forest gardening ... food forests.

Food forests is a sustainable and low-maintenance way to grow food in a system based on the ecosystem of a forest. So instead of having a vegetable bed or a field, you have a forest.

Food forestry fits under the umbrella of permaculture, because you're creating a sustainable, natural system of plants to grow food, except you're concentrating mainly on one ecosystem – the forest.

So what is permaculture anyway?

In the 1970s, the concept of the food forest garden was revived by Bill Mollison and David Holmgren as permaculture – i.e. permanent agriculture. Mollison was working in the Tasmanian rain forests, watching the animals foraging for food, and from the "life-giving abundance and rich interconnectedness of this eco-system" realized that we as humans could grow food in the same way. He also saw how the industrial method of agriculture was destroying soil, poisoning birds and fish and other creatures, and reducing biodiversity. So he and Holmgren began working on a more sustainable agricultural system that works *with* the land and environment, instead of against it.

Food forests are a component of permaculture design. These forests mimic the structure and function of a natural forest ecosystem. They typically have several layers of plants, including tall trees, understory trees, shrubs, herbaceous plants, groundcovers, and root crops. Each layer serves a different function in the ecosystem, such as providing shade, fixing nitrogen, or attracting beneficial insects.

As a side note, you could do the same with a prairie ecosystem or a desert ... though I'm not sure what kind of label we could give the latter. A food desert is a place where people lack access to affordable and nutritious food, so, "Start Your Own Food Desert!" is a book title that would definitely backfire.

At any rate, a successful food forest is one where we let the plants guide us, where we find the systems that allows them to flourish with as little damage as possible to the environment (and hopefully a whole lot of benefit to it).

Proper natural gardening creates a symbiotic relationship between nature and humans – where we all benefit. And that's good stuff.

Forest Gardening History (Kind Of)

The concept of a food forest garden has its roots in traditional agricultural practices, such as agroforestry, or silviculture, which involves growing crops in combination with trees.

In Native societies, tree crops were a vital part of agriculture, providing food, fuel, and building materials. However, the industrial agriculture and monoculture farming practices that were applied on the land when the European colonists rolled in didn't work as well in America.

There always seems to be a chapter in history books in school about the plow being a great invention. On one hand, they're not wrong – but only up to a point. But you know what else is a good invention? Knowing when not to plow, and using other systems to grow crops! Which is what this book is about.

Settler farmers didn't realize that when you're in a new environment, you should use farming methods that work with the local climate and conditions. They cleared forests and plowed slopes, and the rich topsoil that had been held in place by forest roots was all eroded away.

On the prairies, they broke up the virgin topsoil – deep, rich soil that had been made rich by years of burning and natural mulching – and farmed the dickens out of it without doing anything to enrich the soil, until the farmland had nothing left to give.

Native grasses had deep roots (some reaching 6 to 10 feet underground) that held the soil firmly in place to prevent wind erosion. These roots also helped to trap moisture, even during severe drought.

The prairies were plowed into near-oblivion, and severely overgrazed by livestock. The soil, robbed of the deep roots holding it in place, was eroded away. Severe drought exacerbated the problem, and soon windstorms were sending tons of soil aloft. This resulted in the Dust Bowl and 100 million acres of farmland lost.

This isn't to disparage the settler farmers. All my ancestors were settlers. (I've been told that we have Native ancestry, but extensive family trees and DNA results have made it clear that I'm whiter than Miracle Whip.) Most of the folks who came over here to farm the land used the methods that always worked at home, or they looked around and did what the other folks were doing.

Times were tough, and folks just did their best to get by. My great-great grandparents, Eliza and James H. Smith, along with Eliza's mom, Sabrina (Gump) Boyd, farmed the Nodaway River bottoms. The soil was rich there, and the ground was flat, but they occasionally would see all their new corn being taken out by floodwaters.

—The Missouri river is full and still rising—it holds the Nodaway up. Mrs. Boyd cultivated corn last year and this year and no crop yet, but she says, "I have never the righteous forsaken nor their seed begging bread." If they don't raise any crop she is not afaid of starving, for Link Rhodes has plenty of corn. The water got over J. H. Smith's corn on the home farm and he had wheat on his other farm. Mr. Kee on the Bullock place, and Boyd, of the Traub farm, could run boats out on their cornfield. It looks rather discouraging to people that have worked hard and have the water come twice, in o e crop season, and if the water stays up all will be lost; if it goes down soon perhaps there will be some corn left.

JUNO.

Holt County (Mo.) Sentinel, 11 September 1903

Over a hundred years later, farmers still plant on the river bottoms, which is fine until the Missouri River has yet another 500-year flood (we've had three of these since 1993) and dumps a gigantic sandbar across their soybean fields.

But what if the river rats worked *with* the river, and not in spite of it? What if, after the 1952 flood, the Corps of Engineers had not straightened the river? What if they'd

9

created a mitigation zone planted with trees inside the levees, something that matched the old marshes we used to have on the river bottoms?

The Missouri River in 1892-1895, meandering like crazy with lots of islands, marshes, and forests. James, Eliza, and Sabina lived north of Dallas Bend along the Nodaway River (cut off at the top of the image)

The same approximate stretch of the Missouri River now, after the Corps of Engineers straightened the hell out of it and farmers drained the marshes and cut down every tree and brushrow in existence.

What if some of the floodplain farmers bucked tradition and planted rice or some crop that is fine with sitting in a flooded field in June? Maybe they'd get some hell from other farmers at the sale barn and the feed store. On the other hand, they'd be bringing in a crop while everybody else is replanting and hoping their crops mature before it freezes in fall.

A lot of variables go into the farmer's choice of crops for the year. But why can't working with the environment be a variable as well?

If you know that this section of field is swampy and is bad for corn, remember the line, "Insanity is doing the same thing over and over and expecting different results." Talk to your county extension agent about planting a different crop there, or sign up for the U.S. Department of Agriculture's CRP program. It might help the bottom line at year's end.

This is a long, roundabout way to say that it's a good idea to work with the natural systems you've got instead of having to fight them. Going with the flow also saves you a lot of energy.

GETTING STARTED

Food forests mimic nature. A wild or even semi-wild space might look like outright anarchy to most folks who prefer to set their plants in neat rows where each plant never touches another one. A forest garden can be pretty tidy, but for some folks, it might look like the Great Rumpus for plants. Vines leaning on trees, shrubs touching each other, little plants growing underneath big plants, and everybody horning in on each other's space. It's going to be great.

And if you want things to be tidy, that's perfectly fine as well. This is your food forest – so build what feels comfortable and welcoming to you.

In the upcoming chapters, we will delve into the principles and practices that make food forests thrive. We'll explore the natural polycultures and ecological communities that live in your local forests, and use them as a springboard to build your backyard forest.

But this book isn't just about theory – you'll also get a little practical advice on how to design, plant, and maintain your very own food forest garden. This should be a fun journey of creating a sustainable and abundant oasis in your backyard. If it's not fun, then what's the point!

Why Do You Want a Food Forest?

Start by thinking about why you want to grow a food forest. Do you mainly want to grow food for your family? Create a wildlife habitat? Improve the soil quality of your land? Fight a little against climate change? All of the above? Knowing what you want from the garden will help you to focus on those aspects as you're putting all the pieces together.

Perhaps you want to turn an old orchard into something more, or you have a forest plot and want to do cool things with it.

Maybe you want a forest that looks like a forest because you miss the country, and you need a space that gives the heart-uplift of a forest. Perhaps you'd like something more parklike and formal that you can walk into and clear your mind – a place of beauty.

Maybe you want to take that step toward self-sufficiency before the apocalypse rolls up. Or you want an ecologically friendly living space where wildlife can hang out.

Or you want to one-up your neighbors over how ecologically aware you are! (Actually ... don't do this one.)

Avoid Analysis Paralysis

Starting a new garden, especially a food forest, can be overwhelming. There are so many moving parts you need to consider. What kinds of plants will work here? What kind of garden layout is best for my land? What plants are available and what plants can I afford? What if my garden FAILS?!?

I've been doing horticulture stuff since I was in high school and even now, so many years later, that last worry still torpedoes me every time.

It's easy to get caught up in planning and dreaming (or stressing) about all the projects you want to do, but it's important to remember that it's okay to start small.

You might even put off major improvements for the second year, so you don't install a bunch of hardscape and then realize you probably should have installed it ten feet to the left.

Focus on planting one portion of your yard, and once that garden is set up, move on to the next part of your yard. Expect to make mistakes now and then. Even when you make a detailed plan, you're going to find that your plants are going to teach you more about your lot – specifically, how much sunlight they actually get throughout the day (compared to how much light you expected them to get).

Or you might run into other surprises, such as, "My black walnut tree strongly discourages other plants under it branches," or, "I planted a tree not knowing there was a sinkhole underneath and now my poor tree is only three feet tall," or "Wow, the deer ate everything I planted AGAIN."

It's okay if things don't go according to plan. Well, not okay, especially with deer. But don't beat yourself up over it.

Don't be afraid to change your plans as you go. Be flexible. Think of your garden plan as a map. You know where you need to go, but then all these side quests happen, so you'll take a few detours. Sometimes where you end up in a place that wasn't even in the plan in the first place, a place you like better. Gardening is a lot like life in general, really.

A garden is a living thing that evolves over time. You don't need to have everything figured out from the start. The best way to learn is by doing. Plant a few things and see what works. Build on your successes and learn from your failures.

So, don't put too much pressure on yourself to get everything right from the start. It's okay to take your time and learn as you go. Plant what you enjoy and have fun with it. Your food forest will grow and evolve over time, just like you will as a gardener.

Finally, remember that there's no one way to create a food forest. You can modify part of an actual forest or grow your own.

Do you want your forest to look wild and unkempt, just like it is in nature? Go for it!

Would you prefer that it look tidy and neat and have lots of flowers blooming so you can show it to the Garden Club? Sounds like a winner.

It's your garden ... grow it in the way that brings you the most joy and satisfaction.

Site Selection and Assessment

The first step in planning your food forest is to choose a suitable site. A food forest requires a location with plenty of sunlight, good drainage, and access to water. The site should also be protected from strong winds and frost pockets. And it should be convenient, so you can get to the site and drag the water hose to it if necessary.

Look at what you already have on the site. For instance, I really want to plant tall trees at the back of my yard, but there's an easement there with power lines overhead. Any tall tree there is just inviting trouble from the local power company. However, the semi-dwarf apple tree growing there is short enough to keep itself (and me) out of trouble!

Before planting, it's important to assess the site to determine its soil quality, pH levels, and nutrient content. This can be done by sending soil samples to a local laboratory for analysis. Based on the results of the analysis, you can adjust the soil pH and add organic matter to improve soil quality and nutrient content.

Magpie in South Dakota

General Considerations

A food forest requires careful consideration of the site, plant selection, soil preparation – and an understanding of what you are prepared to handle.

Walk around your yard or acreage and look at it closely. Take notes as you consider these factors:

1   **Climate:** What is your climate like? Is it hot and dry or cool and wet? What are the average temperatures and rainfall amounts?

2   **Sunlight:** How much sunlight does your site receive? Is it mostly sunny or mostly shaded? Are there any areas that receive partial shade?

3   **Soil**: What is the quality of your soil? Is it sandy, loamy, or clayey? Does it drain well or hold water?

4   **Water**: How much water is available to your site? Do you have access to a reliable water source, or will you need to rely on rainwater harvesting?

5   **Topography**: What is the slope of your land? Is it flat or hilly? Does water tend to pool in certain areas?

Special Considerations for Arid Areas

I'm going to preface this by saying that I know very little about arid areas or desert gardening – I have never even seen a desert up close, though I really want to.

If you live in a desert area, a straight-up forest garden will need to be created in the desert's image – with plants that thrive in the desert's arid conditions. Most forest plants will not fit that description.

At any rate, if you live in an arid area, follow the cues of the climate, adopt using greywater to irrigate your garden, and concentrate on plants that do well in that climate.

Folks in arid and semi-arid areas need to consider things like sodic soil issues (i.e. salt accumulation), focusing on the local biomes and working with those, and concentrating on what you can sustainably grow there, especially when water is scarce. Water collection systems will be crucial, including greywater systems for "fertigation" (i.e. fertilizing/irrigation). And a whole lot more that requires an entirely new vocabulary as well.

Food forests might be out of the question for some of these areas, for obvious reasons. However, the deserts are full of their own food plants that would be fruitful choices.

City Codes, Neighbors, & HOAs

A book of this sort has to, as a matter of course, deal with the dreaded HOA – Home Ownership Association, which seems to be a purely American thing. I just pray you don't live in a place with an HOA, which I consider an affront to the whole concept of owning property and basically an affront to me personally. If you want to have a car in your own front yard – well, okay, if you want to have a car in your front yard you should probably move to the country.

If you live in a populated area, make sure you know what the local laws are. Even small towns might have laws limiting the number of chickens you can own (or ban roosters), or ban wild areas, or grass over a certain height, or weedy areas.

Instead of going straight to the City Hall to talk to them about what you're planning to do, go first to your University Extension agent. They're likely familiar with local ordinances and are more sympathetic to your cause.

Even better, if you can find an urban farming collective in the area, go straight to them. You want these folks in your corner. They have already been trying to push the needle as far as being allowed to do things like replace their lawn with a pollinator garden, or simply grow vegetables in an HOA that doesn't allow its residents to grow vegetables in the yards of the houses that they own (!?!). These folks will guide you, and you can join them in the fight for fair gardening and urban farming practices despite all the nosey Nellies whose only hobby is criticizing everything you do in your yard that you own because you are paying for it with your hard-earned cash. Perhaps she should take up gardening instead of gossiping.

Anyway.

So join forces with your local gardening/urban farming collective. (You'll also be able to trade seedlings and cuttings with these guys.) Learn what you can do and what you can't do under local ordinances. Talk to the folks at City Hall afterward and let them know what you're doing with your yard and explain why you're doing it. That way, if somebody wants to call and complain, you've already given them your side of the story.

If you're friendly with your neighbors, give them some vegetables. If your neighbors are not friendly, build a nice privacy fence with portulaca or barberries or some thorny food plant growing up the sides. And install cameras so if your neighbors decide to spray herbicides over the top, you have them dead to rights.

I spent some time reading the Tree Law board on Reddit but it depressed me so much I had to quit. Tree Law has posts that are all along the lines of, "This jerk neighbor cut down the 200-year old tree in my yard and I want to know what kind of legal recourse I have, since obviously this a**hole had no right to cut down my freaking 200-year old tree that was IN MY YARD." It's incredibly depressing how many entitled people there are in the world and how many people are just anti-tree in general, or anti-wild spaces, or anti-disheveled spaces.

I feel very lucky to have my friendly neighbors who don't mind my gardening mess, and don't call the police when I go out after dark to stand on a ladder and gather June bugs off my apple tree for my chickens. Thank goodness, or else I'd be in a pickle.

CREATING A LOW-MAINTENANCE FOOD FOREST

At its core, a food forest is a low-maintenance, self-sustaining garden that can produce a variety of fruits, nuts, vegetables, and herbs for years. By planting a food forest, you're not only creating a beautiful and diverse landscape, but you're also supporting local ecosystems and contributing to a more sustainable food system.

Protip: Low maintenance doesn't mean zero maintenance, especially when you're starting out. It will take time and effort to get to a low-energy garden. And practice, too. Sometimes a method you read about isn't the best for your location, but the only way you find out is by experimenting.

Instead of saying, "I made a mistake," you say "This is a learning opportunity" – it's a better way to look at things. Mistakes are bad and evil! But learning opportunities lead to good things.

Do-nothing gardening doesn't mean "abandon it." Completely ignoring your plants means you are abandoning them to the processes of nature and natural succession. Sometimes plants can survive but they usually don't, especially if they've needed watering or fertilizing or cultivation in the past.

If you've been using chemicals on your plants, and you suddenly stop, it will take a little time for the landscape to adjust and the small creatures to return. The plants, which were weakened by their dependence on chemicals, may be more susceptible to diseases and insects for a while. Find other ways to keep the pests down – blast them with a water hose, squish them – and give predatory insects and birds and creatures time to come back and repopulate your yard.

The nice thing about natural farming is that you're working within a system that has been working for millennia, one that's developed itself and runs smoothly with no interference from humans. It has ups and downs, hills and valleys, but it can keep itself rolling despite us – or in spite of us.

Nature has worked since the days of the plow and was thrown off balance in these days of thousand-acre lawns and chemical devastation of all living things in those areas. Different predators will go after your invasive insects. Sometimes frogs will eat the insects, other times the ladybug larvae will do the job, and sometimes birds will pick up the slack. Spiders can do some of the work and their webs are pretty in the morning dew. Though when they build their webs in front of the back door and you walk into them, you can do 15 minutes of cardio in fifteen seconds.

No-Till Gardening

Going to natural farming includes no tilling, since one doesn't see little plows tilling the soil in nature. Seeds fall into the leaf litter or the dried grasses, rains press them against the soil, or into a place covered by mulch, and the root delves into the soil and a little shoot comes up. Grow a thick layer of clover as mulch, and it also puts nitrogen into the soil.

By choosing edible native plants, you're creating a garden that is adapted to your local climate and soil conditions, and you're supporting local ecosystems and contributing to the preservation of native plant species. For instance, the Denver Botanic Gardens in Colorado has a food forest that features a range of native plants, including chokecherry, wild plum, and serviceberry. These plants not only provide food for visitors, but they also support local wildlife, including birds and insects.

When choosing plants for your food forest, it's important to consider not only the edible qualities of the plant but also its role in the ecosystem. For example, some plants might provide food for pollinators, while others might provide habitat for beneficial insects or birds. By designing your food forest with these considerations in mind, you can create a garden that is not only productive but also supports a range of wildlife.

Chickens confronted by the fruit of their labor.

Strategic Total Utter Neglect

The STUN (Strategic Total Utter Neglect) approach to permaculture is all about mimicking nature and letting the natural selection process take its course. In terms of food forest design, this means planting a wide variety of plants and seeing which ones thrive in your specific environment.

Mark Shepherd, a well-known permaculture farmer, advocates for this approach in his own farming practices. He plants a diverse array of crops and varieties and observes how they grow over time. He doesn't view plants that don't survive as failures, but rather as a natural part of the selection process.

Applying this approach to food forests means being open to experimentation and not striving for perfection. Instead of trying to force certain plants to grow in a specific area, the STUN approach allows for plants to naturally establish themselves and grow where they are best suited.

Of course, this doesn't mean completely neglecting your food forest. Regular observation and management are still important. If a plant isn't thriving, it's important to remove it and find a more suitable replacement. The plants that are removed can be used for compost or mulch, further supporting the growth of the remaining plants.

The STUN approach may not be a direct answer to food forest design, but it does provide a valuable perspective on letting nature take its course. Permaculture is not an exact science, and trial and error is a feature, not a bug. By embracing this approach, we can create more resilient and sustainable food forests that work with nature instead of against it.

Chantarelle mushrooms growing on Little Round Top,
Gettysburg Nat'l Battlefield

LEARNIN' ABOUT FORESTS!

If you want to start a food forest, one of the best things you can do is start wandering around in your local forests or stroll down a country trail. Look for food plants. Learn about what's in season – what you can find in the wild.

Pick raspberries, find wild asparagus, gather walnuts, and find morels (if you can!) in the woods. You can find pawpaws or persimmons, gooseberry bushes, ramps, pecan trees, black cherry, crabapples, mayapples, chantarelle mushrooms, and all kinds of other delights.

Waste areas or meadows feature wild garlic, wild mustard, serviceberry, ground cherries, asparagus, strawberries, hazelnuts, amaranth, and leafy greens. Pay attention to where you find these plants, too, so you can understand their light and soil requirements – and replicate them in your yard.

Walking out in nature is a great way to see exactly what kinds of plants do well in the wild, in your climate, before you start setting up your garden.

When you're walking in the woods and you see a bit of forsythia blooming, or a few clumps of daffodils, or bridal wreath spirea sitting there looking all lonesome, then you know that at one time or another, somebody had built a house there. After all, those domesticated plants didn't plant themselves.

Take some time to look around the site, because you might find some food plants that are still growing. Maybe there's a horseradish plant there, or some fennel, or garlic that's grown wild, or an old apple tree. Plants can live for a long time if you don't do something dumb like run over them with a bulldozer.

Once you see these plants in the wild, bearing buckets of food for you without any care whatsoever, you start to understand that you can raise these plants in the same way. You don't have to weed them all the time (but keep the weeds down a little so you can get in and out of the patch) – and you don't have to do a lot of fussy work with these to keep them alive.

So make a special note of the food plants you find growing wild. Maybe the raspberry and blackberry plants in thick clusters under the trees on a country road. A stand of rhubarb still thriving where your great-grandma planted it. Or an asparagus stand growing out by the woodlot.

What you're doing is finding food plants that can, once established, thrive on neglect. You want plants that can take care of themselves. Then you can set up a food forest that doesn't involve a lot of work – which is my favorite part of all this, naturally.

The Layers of the Forest

Using all these different layers of the forest will help increase yields and also capture more energy from the sun and the ground. And you get birds!

To try and keep things simple, here are the seven layers.

1. **Subsurface roots and tubers.** These include the useful plant parts growing underground, such as tubers and roots – carrots, beets, potatoes. Mycelium and hyphae also grow here, the largest part of the fungus. The mushrooms you see are only the fruiting bodies to distribute spores above ground. The actual fungus is underground.
2. **Ground covers.** These are low-growing plants that carpet the ground, like wild ginger.
3. **Ground layer plants.** These are your regularly sized plants that grow above ground, the ones you usually think of when you're growing a vegetable garden.
4. **Brambles and shrubs.** These are taller, woody plants – not big enough to be trees, but big enough to cover themselves with berries.
5. **Vines.** Like grape vines (or poison ivy!!).

6. **Small trees.** These are the shade-friendly trees live all their lives below the high canopy, including pawpaw trees, wild plums, or redbuds.
7. **Large trees.** These reach for the sky and create the upper canopy of the forest. Oaks, hickories, and Chinese chestnuts create these in deciduous forests while the tall evergreens will create the upper canopy in northern coniferous forests.

Your run-of-the-mill vegetable garden generally consists of only two layers – the roots and tubers, then the ground layer. But with a forest setup, you're playing 3D chess with seven whole layers. So technically you're taking vertical gardening to a whole new level!

Guilds: Building Diversity in Your Forest

According to permaculturists, plant guild is a group of plants that work together to create a mutually beneficial ecosystem. This concept was introduced by Bill Mollison in his book *Permaculture: A Designer's Manual*.

In ecology, a guild actually means the opposite – it's a group of closely-related species that exploit, or are competing for, the same resources. So if you planted an ecological guild, they'd all be fighting for the same resources, like a bunch of cousins at a buffet trying to eat all the snow crab legs and not leaving any for the rest of the family.

But in permaculture, guilds act more like Dungeon and Dragon guilds – a group of professionals who team up to protect their own interests and grow together and help each other out, for the most part. They're a mutual-support network.

When it comes to creating a food forest, we can use these natural guilds as inspiration for our own designs. By selecting a diverse group of plants that have complementary needs and functions, we can create a self-sustaining ecosystem that produces an abundance of food. For example, we might choose a nitrogen-fixing plant to improve the soil quality, and pair it with a plant that requires a lot of nitrogen to thrive. This way, the nitrogen-fixing plant provides nutrients to the second plant, and in turn, the second plant provides shade and protection for the first plant.

Ephemerals

Using the guild roles model, we try to pick some plants to perform each of these important jobs, so that we don't have to do them ourselves.

Spring ephemerals are those early wildflowers that pop up in April and May. They play an important part of a forest ecosystem by catching and recycling nutrients that would have otherwise been lost.

How's that work? Well, when plants and leaves die in the fall, they break down through the winter, releasing nutrients into the soil. One of these elements is nitrogen, which is great for leaf greening, but it leaches very quickly out of the soil.

Spring Beauty (Claytonia virginica) and Violets (Viola spp.) are two common ephemeral wildflowers that are lovely and edible.

Over the winter, the ephemerals are slowly growing even under the snow, and their shallow surface roots are soaking up the nutrients from the dead plants, including nitrogen. When the spring thaw starts, these plucky little plants start sprouting leaves in temperatures and conditions that would make any other plant turn black and wilty.

Once the temperatures start to warm up and the canopy overhead fills up with leaves and shades the forest floor, the ephemerals set seed and go dormant, releasing those nutrients back so that other plants can use them.

Ephemerals also include henbit, dead nettle, chickweed, shepherd's purse, wild cress, and many other edible spring greens that you can eat. So you can fill a valuable space in the ecosystem *and* make a little side dish for supper.

The Importance of Diversity to a Thriving Ecosystem

A healthy food forest is a diverse food forest. By including a variety of plant species, you can help to support a range of pollinators and beneficial insects, reduce the risk of pest outbreaks, and improve soil health. As your food forest grows and matures, keep adding in a mix of fruit and nut trees, berry bushes, herbs, flowers, and even mushrooms, to create a diverse and thriving ecosystem.

But why is diversity important here?

Grandma Anna told me stories about growing up during the Great Depression. Her dad, George Salmen, was a farmer in North Dakota, and his farm had a wide range of crops, not just wheat, along with livestock and a vegetable garden, so she said they were able to do well during those years because he had a diversity of crops and food to fall back on. If one crop didn't do well, the other crops could pick up the slack. If disease killed the wheat, he still had corn.

In the wider scope of things, diversity also serves another purpose. Nature hates a monoculture. If you have a big field of one variety of corn, just about every plant in that field will be susceptible to the exact same diseases and attacked by the same pests. If Japanese beetles attack one plant, they'll attack the rest of the plants, and then you're in trouble at harvest time.

Natural ecosystems, particularly forests, show less fragility and more resilience than agricultural systems. They tend to develop high species and genetic diversity and a high degree of functional interconnection as they mature.... Natural systems require little or no management. They create little or no waste or pollution. We hope to gain all these advantages by mimicking forest ecosystems.

--*Edible Forest Gardens, Vol. 1: Ecological Vision and Theory for Temperate Climate Permaculture.* Dave Jacke with Eric Toensmeier

This brings us to another troubling point. Our national and local forests have turned into small patches, fragmented islands that have trouble maintaining high diversity and stable populations. To combat this, we need to connect these little islands of forest in corridor that link them to nearby natural habitats.

"Will your forest garden be an island or part of a web?" *Edible Forest Gardens, Vol. 1*

Nature hates a monoculture -- and hungry insects and diseases love them. In a monoculture, it's easy for a disease or hungry insect and all its progeny to skip from one plant to the next. If plants are scattered through the landscape as nature intended, then it's easy for plants to keep from being overrun by these pests ... they have some defense.

Young forests have trouble with being overrun by aphids or other pest insects because beneficial insects haven't gotten established yet. However, in older forests, the insect population has had time to balance itself out with pests and beneficial insects, and the number and diversity of species is much higher. the older forest likely has more niches for all these disparate species, and they've had more time to come in, especially if resources are plentiful. This kind of functional diversity creates a better place for elements to interconnect and work more effectively together.

Diversity isn't just a question of monocultures – we also suffer from a lack of diversity in our plant gene pools. Back in 1970, a disease called southern lead blight decimated nearly half of the corn in the U.S. At the time, most of the corn varieties all shared a parent plant – and from this parent they'd inherited a trait that made them susceptible to the blight. This is a common thing among commercially-grown crops – the gene pool for these has been narrowed substantially.

Back during the turn of the century, we had all kinds of different varieties of corn and onions and flowers and apples, but when hybrids appeared in the market, many of those different varieties were dropped from commerce, and some ended up vanishing altogether.

The surge in heirloom plants – plants that have been grown for centuries – is the answer to this shallow gene pool. And you can save the seeds with heirlooms and grow them again next year.

Here's another story about the importance of having a lot of variety, i.e. a big gene pool, in plants.

The folks in South American have tons of different types of potatoes in all kinds of different colors and shapes and sizes, so if several potato varieties are susceptible to blight, well, no sweat, there are plenty of others that aren't. The folks in Ireland had only one type of potato – and that potato succumbed to the blight that led to the Potato Famine.

Potato infested with **Phytophthora infestans.**
Credit: *Agricultural Research Service.*

But remember that it wasn't simply the potato blight that killed the Irish. Their British absentee landlords forced them off their homes and lands, took all their wheat and grains, and left the Irish people to try to survive only on the meager crop of potatoes they were raising.

As John Mitchel pointed out, "The English, indeed, call the famine a 'dispensation of Providence;' and ascribe it entirely to the blight on potatoes. But potatoes failed in like manner all over Europe, yet there was no famine save in Ireland. The British account of the matter, then, is first, a fraud; second, a blasphemy. The Almighty, indeed, sent the potato blight, but the English created the Famine."

In a monoculture, if your one crop fails, then you're out of luck. In a forest garden with a diversity of crops, if one fails, you're still going to have a steady supply of food.

*

Individual yields for some plants may be lower when grown in a food forest. Raspberries growing along the forest's edge won't yield as much as those grown in full sun. Indeed, the raspberries that are growing under my mulberry tree are not yielding as much as the ones I have cut back and growing in full sun.

But yield is not everything! These raspberries that are cuddled under the mulberry trees with violets around their feet are a haven for beneficial insects and provide habitat for wildlife. Even better, these raspberries are extending the season. My shade raspberries are ripening a week or two later than the sun raspberries, which brings me much joy.

And sometimes you hit upon a combination of plants that live happily together, benefiting each other. Then your garden is golden.

A diverse setup gives you some wiggle room and alternatives for when things go sideways – because you know they always will. A diverse system is a stable system.

Now we're going to talk about something cool that I ran across that I think would work pretty well for smaller areas. And if you're crazy about forests in general, you can adapt this in a million ways and in a million places and it would be pretty awesome. For the purposes of this book, I'm going to talk about microforests and how we can apply these to food forests.

A Miyawaki planting with 3,300 trees at Hanji Farms, Belagavi, Karnataka, India, on 15 April 2017. Photo courtesy of Green Saviours, who had planted four microforests when this picture was taken. They followed this up with a fifth microforest with 5,000 saplings!

MICROFORESTS
The Power of Small-Scale Systems

While writing this book, I've been thinking about ways to bring more food shrubs and trees to my yard and make a cool little forest, carefully planning and mentally moving semi-dwarf fruit trees around my yard. *I should plant a white peach tree here by the chicky pen ... no, the squirrels will be in the branches in two seconds if it's here, maybe move it to the other side of the shed ...*

But then I ran across Akira Miyawaki's microforest work, and I'm all over this.

I want a forest. I moved out of Nodaway in late 1994 and have been pining for its forests ever since. It's not like they're a million miles away, but I just want to be able to walk out my back door and vanish into the woods like I used to back in the old days.

And now I'm reading about the Miyawaki method of creating tiny forests, aka pocket forests, aka mini-forests, and I'm so ready to put a variation of it in my backyard.

A tiny forest is a fast-growing and dense stand of native trees that is usually about the size of a tennis court, though you can make it any size you want.

If you remember your science classes, a new patch of trees will compete as they grow up. Among trees, it's a race to the top, because the tallest trees will get the most sunlight. So while a tree you plant all by itself in the front yard will take its sweet time in growing, the same tree when planted with a bunch of cohorts around it is going to start stretching!

So these forests can sometimes mature within ten years, instead of taking 35 years or more to reach maturity.

The microforest really appeals to me because my whole philosophy is about starting small so I can keep up with the work. It's easier to add something on later – and it's hard to maintain something that you can't keep up with. So I'm all over this idea.

Now I'm going to explain how Miyawaki envisioned these forests. Then I'm going to talk about making a food forest version of these, which would have several key differences, depending on what you choose to go with.

Planting a Miyawaki forest at Kanakakunnu Palace, Thiruvananthapuram, Kerala. Note the chalk grid. Due to the dry soil, they tilled in coconut coir to hold water.

The Miyawaki forest – nine months later! It has flourished, and now honeybees, owls, and many butterflies and birds live here. Now they're planting 22 more microforests! Read more: https://www.crowdforesting.org/miyawaki-forest/kanakakkunnu-wild-forest

What is a Miyawaki Forest?

Akira Miyawaki was a botanist who studied primal forests in Japan. Though forests were hard to find there, there were always untouched forests, centuries old, around temples and shrines. He studied the plants and trees that made these and found that these ancient native species were able to stand up much better against natural disasters.

He began planting miniature forests of native trees – the specific natives found in that specific area – around Japan. He collected seeds and cuttings from these plants, raised their saplings, then planted them close together. This resulted in swiftly-growing, dense forests of native species that could withstand anything that the world could throw at them.

Miyawaki chose these trees because they'd been growing for centuries, untended by man, withstanding everything that nature had thrown at them for all those years. Anything that was this self-sufficient was the ultimate no-maintenance plant, and that's what he wanted to grow!

Naturally, you have to care for the plants until they get established – so you mulch them, cut back the weeds, and water them when it's dry. After three years they should be completely self-sufficient. Any plant that needed maintenance after five years wasn't worth it – those plants could go away and wilt.

The tiny forest as Miyawaki envisioned it was meant to be grown only with trees native to your area. The proper

41

study of this would involve visiting old native forests that have been unchanged by man for centuries.

For many people, this might not be possible.

I live in Missouri, which sports 14 million acres of forestland, thanks to the Ozarks down south. So we should have a ton of old-growth forest, right? Sorry. Here are the stats from the Missouri Department of Conservation.

Out of those 14 million acres of forest, Missouri has:

- 62,000 acres of forest with a few trees that are 130 years old.
- Fewer than 8,000 acres of relatively undisturbed old-growth forest.
- Only 800 acres of excellent examples of old-growth forest.
- That works out to **less than 0.005 of 1% of Missouri's forestland.**

Not near enough!

Since we don't have much in the way of mature forests, the next best thing would be to consult historical writings. For the area where I live, that would involve reading diaries of settlers and explorers who lived here or passed through, such as Lewis and Clark.

Your best bet would be to find narratives for the Native tribes that lived in the area before the settlers – in my case, the Otoe-Missouria, the Kansa, Kickapoo, Osage, and Sioux Nation – before settlers drastically changed the landscape.

It's not easy to find these narratives since many of these records and stories have been erased and purposefully lost to time by the settlers. An appalling amount of knowledge and history has been destroyed over hundreds of years, and it's still being done. But if you can reclaim a piece of history and understand the land (and the people on it) the way it was in the olden days, that's worthwhile work.

At any rate, Miyawaki said that this deep understanding of the land and the historically native plants and trees is vital to planting a microforest – that a microforest should be made up of true native trees that can prosper in your climate, just as they have for centuries.

You can make the microforest any shape you like. Just make sure that the forest area is at least 12 feet wide with no interruptions (such as a path). Keep the path at least 12 feet away from the edge of the forest, and plant thorny plants away from the path so they don't snag you. Make the path at least 20 inches wide so you can walk along it with a wheelbarrow without crashing into stuff.

In a true microforest, you'd plan on having about three sapling trees per yard/meter. Growing the saplings so close together will make them grow faster – giving you a leafy canopy overhead in only about three years!

Plant the trees at random, just as they would be in nature, not in straight rows. Miyawaki suggested that, for true randomness, have a child place the trees where they are to be planted.

Don't plant trees from the same species next to each other. Make your microforest like those fancy state dinners – mix up the guests and make them sit next to different people. Give each sapling at least one foot of space.

So that is the Miyawaki ideal for a microforest.

However! I will add a caveat here.

CAVEAT:

IT'S OKAY TO DO THINGS YOUR OWN WAY.

Embroider this on a napkin and wave it like a flag every time a stickler starts talking about how things are *supposed* to be.

A microforest doesn't have to be perfectly native. If you plant a microforest with some non-native trees, it will still be awesome, and Miyawaki's ghost will not haunt your horticultural dreams.

Our #1 goal here is to plant a crapload of diverse trees that will grow into a forest that will take care of itself. Right now we need forests, and quickly, to cool urban heat islands, sequester carbon, help pollinators and wildlife, create biodiversity, along with many other reasons.

But for the purposes of this book, we also want a little forest that will shower us with tasty stuff. And not all of us have a lot of space – maybe just a backyard's worth of land to work with.

Now I'm going to take a short detour into orchard management.

QUINCUNX SYSTEM.

Orchard Growers Love This One Simple Trick!

Did you know that orchard growers have a simple trick that resembles Miyawaki's idea?

Orchard growers sometimes use a planting pattern called the quincunx to increase yields by filling in extra space while the young trees are maturing.

Quincunx is from the Latin for five-twelfths, and it's basically when you arrange four objects in a square and place the fifth in the middle. It allows you to plant more trees in a smaller area, and these trees bear more quickly.

However, once the canopy starts filling in, the orchard manager needs to thin out the central tree in each quincunx in order to keep the main trees bearing properly – a hard thing to do when each central tree is healthy and productive.

There are numerous systems for planting orchards. Many fruit growers have worked out their own systems. Those mostly used are the rectangular, quincunx, alternate and hexagonal, or modifications of them.

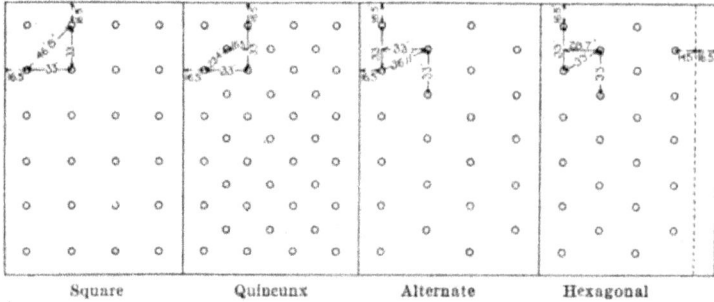

| Square | Quincunx | Alternate | Hexagonal |

Figure 3.—Four basic plans of planting an orchard.

If you'd like to try the quincunx method with your fruit trees for early bearing, knock yourself out. Personally, I couldn't cut down the central tree once the fruit trees begin running into each other, even if that central tree had started as a little seedling whip I bought on clearance at the local farm store for $12. Then again, you know I'm cheap.

The nice thing about domesticated food trees is that they're grafted onto rootstock that keeps their growth within bounds, so you can choose semi-dwarf fruit trees (12 to 15 feet tall) and dwarf trees (8 to 10 feet) in order to squeeze more trees into a small space.

If you're not quite ready to commit to an utterly random landscape plan, planting in a quincunx might be a more pleasing alternative.

That said, let's create a variation of Miyawaki's microforest – a micro food forest.

Applying Miyawaki's Ideas to Your Food Forest

So we're going to do things a little differently. Feel free to choose native food-bearing trees, shrubs, and shade-loving perennials for this miniforest. Miyawaki is right about using plants that have been proven to work for centuries.

But if you want to mix in apple trees and a self-fertile cherry tree and other non-native fruit trees, go ahead. If your fruit trees need 10 or 20 feet of space, give it to them, especially if you paid $70 for these fruit trees.

So a miniforest of dwarf and semidwarf fruit trees should be a little closer together than usual, and in this case it's a good idea to stick with trees that all grow to be about the same height. I wouldn't mix dwarf fruit trees with chestnut or walnut trees, for example, since if wouldn't be long before the nut trees overshadowed the apple trees. Fruit trees are not a forest-dwelling species, so they'd stop bearing fruit in the shade and eventually croak.

However, if you paired the fruit trees with smaller trees such as redbud, hazelnut, pawpaw, you'd be in business. And you can get dwarf and semidwarf nut trees as well! Though these seem to be harder to find, they're certainly worth the search.

So you're getting a bunch of domesticated dwarf and semi-dwarf fruit and nut trees, mixed with native fruit and nut trees that stay fairly short, and you're not planting them within 3 feet of each other – more like 10 to 15 feet. You should put them far enough apart so you can walk around each tree and pick the fruit without getting your hair caught in the branches of the trees behind you.

There's an almond tree that's not a true almond (it's a hybrid of a peach and almond) called 'Reliable' that gets to about 12 feet tall and is self-fertile.

To my absolute surprise, the Allegheny and the Georgiana chinquapins (*Castanea pumila* and *Castanea alnifolia,* respectively) are available as shrubs or dwarf trees (as well as full-sized small trees). More about these later in the book, in the pages about Nut Trees. But some versions of these trees stay very short and bear delicious chestnuts.

*

Forest gardens larger than a half-acre are going to take a lot more time to build up, especially if you're only a crew of one. If this is the case for you, start by building a microforest as the nucleus of your operation. Once it's in good shape, plant another nearby, then another. Then, as your forest expands, you can set up new plantings and trees to merge all of the microforest nuclei together. This process takes longer, but it also saves money and sanity!

Even cooler: Set up microforests to mimic different ecosystem patterns, such as shrubby areas, old growth forest, and thickets. Place these ecosystems where they'd make sense, of course (such as putting boggy plants in boggy areas). Add clearings and gaps between the microforests to let sunlight in.

Enhancing an Already-Existing Forest

Not gonna lie, but I'm jealous of anybody working with an actual forest or even a neglected woodlot. My dream is to be able to live in a place where I can have a forest of my own, however small. So if you have one, enjoy it!

But with existing forest, there are more ecological considerations. An existing forest would likely benefit more when you look at it with an eye toward improving its ecological balance, enhancing what the forest already has, and providing it with what it needs.

One side note: Don't replace real forests with artificial forests. That is, don't go in and modify a real forest until it's unrecognizable as a wild area, as if its only value is in how much food it can bear for you. We need wild spaces that are not here to benefit only humans. A big part of food forests is to grow food, but it's also to create space for insects, birds, and other creatures, especially in this world where everything is commodified.

And not everything can be grown in a food forest. You'll need a garden space for sun-loving plants that won't be happy in partial shade. And that's fine.

Assessing What Your Forest Has to Offer

A forest inventory is a deep exploration of a forest's resources, helping you understand what you have, forest-wise, and identifying what you need to improve upon.

Call your local forester or contact your county University Extension Service for help with this because there's a lot involved, especially if forestry is a new discipline for you. I mean, I'm a horticulturist and a master naturalist, and forest surveys are still daunting for me.

Tread carefully as you modify your forest. Start by choosing the path of minimal change. Weed out invasive non-native plants, add woodland perennials that are native to your area, tidy up the bramble patches. Then add shrubs and shade-tolerant trees if you're ambitious.

You can open up a gap in the forest to bring in more light, creating a new planting in that area.

If you are setting up your forest garden around older trees, you'll need to select plants that can grow well in partial to heavy shade.

Start by looking at the existing vegetation under your trees to see how much light is reaching the ground. If the shade is so thick that grass can't grow underneath, you'll have to cut off some bottom limbs to let some light in – or you could plant mushrooms and make that area into a cool place where you can escape the summer heat. And right now while we're having the hottest days ever in all of recorded history, all over the world, we could use a little extra shade right now.

I knew climate change was getting out of hand, but this is ridiculous...

The Forest's Edge: Where Sun & Shadow Meet

The edge of the woods is where a great deal of diversity happens (at least in natural conditions). It's the area of thickest growth as shade and sun plants both vie for sunlight. In this area of shade and light, you can mix all kinds of plants.

51

Heavy Shade

Around the edges of the deep shade trees, you could grow smaller trees, then outside of those, sun-loving plants such as brambles and berry bushes.

Under deciduous trees, grow early spring cole crops that mature before the trees fully leaf out.

In early spring, scatter radish seeds and lettuce seeds under the bare-limbed trees, where they will grow until the leaves come out. Once the leaves shade the ground, plant shade-friendly summer greens. If you're not sure whether a specific variety of greens can handle the shade, just put a couple of them in and see if they grow or not.

It's good to experiment to see what your plants can handle and what they can't. Sometimes they disappoint. Sometimes they'll surprise you!

Heavy shade will limit with what you can plant. A number of spring ephemerals – the pretty wildflowers that pop up in early spring and fade away by summer – will do well in shade. Flowers like trout lilies, toothwort, wild violets, and ramps (*Allium tricoccum* for American ramps and *A. unsinum* for European ramps) are pretty and edible. They also bring in early bees and other pollinators that are looking up and down and all around for nectar.

Don't leave the soil bare, even in shade. Even if you're just growing a ground cover like dead nettle (*Lamium* spp.) or wild ginger (*Asarum canadense*), growing plants on bare soil is good for the soil from an ecological standpoint. Dead nettle is also edible (but this plant is not related to stinging nettle, nor does it sting. *Lamium* is a member of the mint family.)

Windbreak next to nesting cover and small grain food source provides diverse upland bird wildlife habitat near Geraldine, MT. Photo courtesy of USDA-NRCS Montana.

Windbreaks

These seem to be a relic of a bygone age as many farmers in this area seem intent on cutting down every stand of trees on their acreage in order to "increase yields."

Windbreaks are a godsend if you live in an especially windy area. These block the strong prevailing winds and snow, creating a calm area around the house and barn that absorbs the sun's warmth in winter, and they keep the drifting snow off your roads.

You can add nut trees and certain berry shrubs to the windbreaks that will hold the drifting snow out of your homestead and then in summer and fall provide food. Plant food plants for area wildlife. For instance, Rocky Mountain junipers are a good windbreak tree for western folks, and provide juniper berries for livestock and wildlife such as wild turkeys.

If you have an old windbreak that needs to have dead and dying trees removed, or if you live in a place where you need a windbreak, contact the USDA-NRCS about their Windbreak Restoration program.

Creating Sun Traps in Your Food Forest

When designing a food forest, one of the key considerations is how to maximize sunlight for your sun-loving plants. In temperate climates, a common strategy is to create a "sun trap" in the center of the garden. The forest has a horseshoe-shaped clearing facing south that allows the sun to reach into the forest and shine on these plants, while still retaining some shade from smaller trees and shrubs. This also stays nice and warm, so the sun trap creates a microclimate that is ideal for growing a wide range of edible plants.

One important consideration when designing your sun trap is to choose the right plants to grow in this area. You'll want to choose plants that are well-suited to the amount of sunlight and shade in the area, and that can handle the microclimate created by the sun trap. You can grow your sun-loving herbs and vegetables in the center of the sun trap, and more shade-tolerant plants around the edges.

Ramps going crazy in the woods!

DESIGNING YOUR FOOD FOREST

"People frequently ask how much land they need for self-sufficiency. The answer is, 'as much as you can control.' Any more and you lose self-sufficiency, let alone the ability to produce an excess. If people ask, 'where do I start?" then the answer is always, 'at your doorstep.'" – Bill Mollison

CREATING THE LANDSCAPE PLANS

Now it's time to draw up some rough plans. You can hire a landscape designer to come in and do the work for you – but you'll likely be expected to buy all your plants with her company and have them installed with her landscape crew. Whereas I'm cheap.

If you're part of the DIY crowd but you haven't done much garden design, there is free landscape design software and apps that are available for your computer or phone. You can find these by doing an internet search for "best free landscape design programs," then look through a few online articles to see what's available. The right software program can make this process much more enjoyable.

If you consider the internet to be the bane of all humanity (you're not wrong), you can also use old-fashioned paper, pencils, protractor, and patience.

1   **Start with a rough sketch**: Using graph paper, draw a rough sketch of your site. Indicate the location of any existing structures, such as your house, garage, or shed.

2   **Identify your zones:** Divide your site into zones based on how frequently you will visit each area. Zone 1 is the area closest to your home, where you will plant the most frequently used crops. Zone 2 is slightly further away, where you will plant crops that require less attention. Zone 3 is even further away, where you will plant crops that require minimal attention, such as fruit and nut trees.

3   **Determine the appropriate scale**: Consider the size of your site and the amount of time and resources you have available. It's better to start small and expand over time than to overwhelm yourself (especially when summer arrives and everything needs weeding).

4   **Plan your plantings:** Select a mix of crops that are appropriate for your climate and soil. Consider planting groups of plants that work together to create a self-sustaining ecosystem (aka guilds).

5   **Choose your planting locations:** Group together plants that require similar conditions. All water-loving plants to the boggy part of the yard; all heat-loving plants next to that south-facing brick wall, etc.

6   **Incorporate vertical layers:** Consider the height of your plants and incorporate vertical layers to maximize space and sunlight. Grow vines.

7   **Add supporting elements:** Remember to leave a spot for your compost bins, rainwater harvesting systems, and paths to make your food forest garden more productive and manageable.

8   **Give everything space:** Make your paths and access points wide enough to run a wheelbarrow through. Give your plants a little extra space, too. There have been many times when I thought, "This little plant is not going to get THAT big." Welp! I was wrong.

9   **Side note:** Just because it's a forest doesn't mean everything is in deep dark shade. There are glades with open light, edges, and growing areas that aren't yet shaded. Clearings allow for sun-loving plants. In southern areas, the sun is harsher, so this shade will give plants a cool place to grow when temperatures are blistering.

Additional note: I have a lot of additional garden design information in my book <u>Design of the Times: How to Plan Glorious Landscapes and Gardens</u> if you really want to dig into this aspect of gardening.

Currants growing in a renovated shelterbelt planted in 2010. Producer made jelly and donated it to the local food bank. Hill County, MT, June 2012. Photo courtesy of the USDA-NRCS Montana.

Wild vs. Manicured: Discovering Your Style

There's a continuum that moves from "manicured to an unrealistic degree" to "absolute chaos." A formal garden would be in the manicured realm, while an overgrown weed-tree bramble that you can't even walk through is in the realm of chaos.

Planting all your trees in straight lines would make your food forest a little too formal – a little too unnatural. However, some folks consider this more of a feature than a flaw.

Personally, I lean toward mimicking a real forest – staggered trees, areas of brambles for wildlife (and to grow raspberries!), pretty areas that look a little overgrown and wild. However, unlike a real forest, you can walk through a food forest without getting tangled up in greenbriers or gooseberry thorns, or tripping over logs, or sliding all the way down a long hill and falling into a creek at the bottom. It is a garden, after all, so you are designing it to be beautiful, and easy to enjoy. It's hard to enjoy a garden if various plants *cough*ROSES*cough* keep pouncing on you with their little knives drawn.

If a few trees die in a forest setting, I leave them standing instead of cutting them down, because "snag" trees are good places for wildlife to live.

I have a silver maple tree in my yard that's mostly died, but this year, three red-headed woodpeckers have set up a condominium in it. Red-headed woodpeckers are very grand and fancy. They have a striking white and black on their wings and when they come sailing in, it's very impressive. It's like seeing someone go by in a top hat and a cloak. I might just have the tree people top this maple (since it's dead) and leave enough of the tree so the woodpeckers can keep building.

Designing for Water, Landscape, and Access

Designing a food forest can be a daunting task, especially if you are new to gardening. And sometimes it's daunting even if you're an old hand at gardening! However, by considering three key factors - water, landscape, and access - you can create a thriving and functional food forest that is easy to maintain and provides a bountiful harvest.

Water

Water is essential for any garden, but designing your food forest with water in mind can save you a lot of time and effort in the long run. Before planting, consider where your water sources are and how easy they are to access. If you have to drag watering cans up a flight of stairs, growing vegetables there may not be the best idea. Look for locations that are close to water sources or install a rainwater harvesting system to make watering your garden more convenient and eco-friendly.

Landscape

Understanding the landscape of your garden is crucial when designing your food forest. Observe the sun and wind patterns and take note of areas that receive shade at different times of the day. Plant shade-tolerant crops in areas that are shaded for most of the day, and sun-loving plants in areas that receive full sun. Check for areas that may become waterlogged during rainy periods or drought-prone during the dry season. By working with the natural conditions of your garden, you can create a more resilient food forest that thrives year-round.

Access

Creating easy access to your food forest is essential for ensuring that you can easily tend to your plants and harvest your crops. Consider where the most convenient location for planting would be. Plant herbs and vegetables closest to the kitchen, so that you can easily grab fresh ingredients while cooking. For larger food forests, create paths and zones to easily navigate through the garden. Make sure you have easy access to the compost bin. Consider installing raised garden beds if bending or kneeling is difficult for you.

But also create your gardens with an eye toward building a pretty living space outside. Put a few antique roses into the landscape where you can see and smell them from the kitchen window. Design the area so you can go outside and read a book under a bower of sunflowers, or watch a grackle chasing a squirrel in the mulberry tree, or sit next to a firepit and listen to owls at midnight.

You will need more space than you think. Make your paths wide enough to wheel a heavily-laden wheelbarrow along them.

Paths also gives you more open areas in your garden to grow herbs and other sun-loving plants. Growing thyme and other low-growing herbs to walk on will add fragrance to the garden.

You can lay down paving stones, or you can plant low-growing thyme, chamomile, or white clover in the walkway in a patchwork of creeping plants that are still nice to walk on.

If you are going to choose plants that need more maintenance or work, put them closer to your house so you can easily putter around with them when you have a minute. Low-maintenance plants can be placed farther away. Also, group plants with similar needs together.

If you are going to bring in tons of mulch or compost and spread it around with a tractor, be sure to have a place where these materials can be dumped, and enough space to use equipment larger than a shovel.

Observation

Finally, observe your garden and yourself. What areas are the most neglected? What plants do you find yourself not using? What resources are currently being underutilized? Use this knowledge to improve your garden design and make the most of your available resources.

Start with planting what you like to eat.

Learn through trial and error. Remember, a garden is an evolving system, and it's okay to make changes as you learn what works and what doesn't. Sometimes things don't work the way you planned. Sometimes plants will croak. This happens to every gardener. I'm a professional, and I have probably killed more plants than most of you! It's all part of the learning process.

Use Weeds to Understand Your Terrain

"What does this mean, using weeds to understand your lawn? You're not asking me to talk to them, are you? I mean, that's the quickest way to get a nettle to stab you. They're really touchy about their personal space."

The types of plants growing on your land can help clue you in to whether that particular area has full sun, or shade, or a swampy spot.

Do you have nice, thick grass on your yard, but then it gets a little patchy when it gets close to the fence where there's a tree on the other side? The thick grass indicates full sun, but the patchy parts indicates that there's more shade there that you bargained for.

Lots of moss indicates a cool spot in a very shady place. A group of reeds, cattails, or horsetail plants (*Equisetum* spp.) indicates a damp or swampy part of the yard.

I remembered how ground ivy grew thickly around the front of my Great-grandpa Ben's house, and how shady it was there. That made me realize that the ground ivy that now covers half of my lawn means that the silver maple trees are providing more shade than I realized. So that's why my 'Evelyn' rose refused to grow there!

The plants that are already in your yard can help you get a clearer picture about what your land can provide, and help you correct any wrong notions you might have gotten about it.

GROWING A FOOD FOREST

Harnessing the Power of Swales

If you have a slope, or regular flooding, you should probably have swales. These are basically speed bumps for the water running downhill. They catch the water before it runs off and give it time to soak into the soil, which is a good use of resources.

If you have a lot of land, this will mean a lot of digging. In the worst runoff areas, one way to create a swale would be to stack fallen branches and limbs across the runoff areas and make a **hügelkultur** mound out of it. Stack the branches, cover them with fallen leaves, then compost if you have it, then a layer of soil. This way, you are getting rid of a lot of old brush, creating swales without a ton of digging, and creating fertile soil to boot.

If you've been cutting down large trees and the wood isn't good for burning in a wood stove, use the logs as swales. Don't even cover them with soil – the runoff water will do this naturally. Again, this saves you some digging, especially if you're having to use heavy equipment. After a year or two, you can use the swales for planting.

PLANNING YOUR FOOD FOREST: THE PLANTS

Choosing the Best Edible Plants for Your Region

When selecting plants for your food forest, it's important to choose species that are well adapted to your local climate and soil conditions. Native plants are often the best choice, as they are already adapted to the local environment and require less water and fertilizer than non-native species.

For example, if you live in the Pacific Northwest, you might consider planting native berries, such as huckleberry, salmonberry, and elderberry. These plants are well adapted to the cool, damp climate of the region and provide a source of fresh, nutritious fruit.

Not every plant you grow has to be a native species – many introduced plants will do well in your area.

Here's a good rule of thumb, one that has served me well when I was a horticulturist: If you keep trying to grow a plant and it keeps croaking, stop growing that plant. Somehow that plant isn't getting what it needs, and the plant won't tell you why it stays puny and then fails. I like flax plants and lupines, but they don't do well in this area, where the summers are so hot, so instead of putting in extra work to keep them alive, I just grow other things that prosper – plants that grow themselves. It saves me time and patience.

When it comes to fussy plants, I always quote what E.B. White wrote about his editor wife Katherine S. White: "She never hesitated, she never fussed, and she was quite rough with flowers, as if to say, 'If you can't take the heat, go away somewhere and wilt.'"

Of course, if you absolutely have your heart set on a specific plant, by all means, keep trying. But life is short, so don't finish reading a book you hate and don't keep trying to grow a plant that won't prosper.

Native Plants vs. Non-Native Plants

Native plants are plants that have been growing in a specific area for a long time without human intervention.

But think about that definition because I'm going to throw a wrench into the works. We *assume* that these native plants have always been there.

But before the white man barged onto the continent, Native Americans were traveling far and wide, bringing pots and bags of seeds and cuttings from home, and trading them with other Native Americans, back in the days when they had bustling cities and towns, back when they traveled freely on land and water.

The idea of Native folks taking plants from place to place is no different than the settler women on a wagon train bringing cuttings of her beloved lilac and roses from home, or the woman in steerage coming over from Switzerland carrying seeds in her bag. So perhaps some of the plants that we consider native to an area were brought in from elsewhere by Native Americans, ages ago.

As a long-time volunteer for the Missouri Department of Conservation and a master naturalist, I'm a big fan of native plants. These plants have already lived in your area for ages, so they'll take whatever your local climate will throw at them. That means you don't have to coddle them. If you take the time to get them established, and help them along now and then, you'll have a bunch of plants that make your life easier and look great doing it.

However, there are too many great plants in the world to limit yourself to only natives. By all means, protect native plants and raise them in your yard. But also enjoy the great variety of plants from all over the world, just as long as they're not invasive in your area, and won't cause a problem to your local wildlife.

That doesn't automatically mean I'm anti-native. I'm more of a pragmatist than a purist.

Some native plant purists want only natives in landscapes and nothing else. Even having a mostly native landscape with only a few non-natives isn't good enough.

Being a purist isn't bad. I too have been known to hold Big Opinions on various things, such as the serial comma, or how the word "impacted" can go impact itself out of existence impactfully, and how work meetings are a stupid way to use this one life that we're given.

I want to see native plants used widely and responsibly. Creating a whole new ecosystem in a threatened area is a big deal, and native plants fit the bill so nicely. Say you're rebuilding a marsh in a place that's been farmland for decades. You should absolutely get the right native plants and spend time fixing the habitat so those natives will grow and thrive, and the wildlife does the same. That's awesome work! I want to see more of that!

I also am in favor of having cool plants from elsewhere in your garden to keep things lively. The world is so filled with cool plants that settle in next to natives and behave themselves, and I hate to deprive my garden of something cool. I'm getting some Missouri chestnuts for my yard, but I also want to get some of the *Malus sieversii* apples from Kazakhstan and grow a ton of them in my yard because I've been excited about them for years. They're not native, but dang they look cool.

I also object to a practice called moonscaping that some native landscaping companies will do in trying to establish native populations. They come in and spray herbicides all over, then till up the ground, then spray herbicides again before planting the pure native landscapes.

It's like ... what are you *doing*, man? If you are planting natives, what's with the herbicide spraying? And the erosion that moonscaping causes? And the part where you destroy an ecosystem that's already established?

What happens when that ecosystem is destroyed and the bare earth is exposed to sunlight? Dormant seeds of invasive weeds, seeing sunlight now that the overhead plants have been killed off, can sprout. Now that there are no other established plants to suppress them, the invasive weeds take over the native planting.

So by moonscaping, you're creating a space for the evil opposite of native species to run rampant.

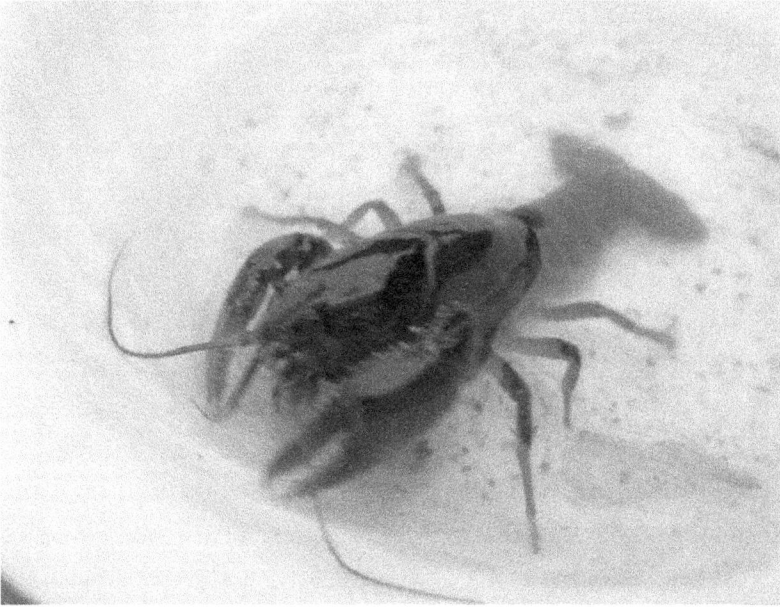

This devil crawdad (Lacunicambarus aff. diogenes) was living in my yard. They dig holes in moist soil down to the water line and hang out there. I understand you can lure them out of their holes with a piece of bacon tied to a string. I keep meaning to try this, but somebody keeps eating the bacon. Actually, that's me.

Sharing the Space with Wildlife

Part of planting a natural garden is sharing it with the wildlife.

In nature, plants and animals have evolved to coexist and support each other. When we walk through a forest, we can see how different species of plants and animals work together to create a balanced ecosystem. For example, trees have roots that reach into the subsoil and bring up nutrients that the little plants can't reach. The blanket of autumn leaves that the tree drops protects the little plants and creatures through winter, and as the leaves break down to nourish the soil, they release those nutrients the trees brought up from the subsoil. Mycorrhizae, a fungus that lives on tree roots, increases the absorption powers of the roots and gives the tree nutrients that would be ordinarily hard to absorb.

Ants live on the trees to protect their sundew-producing aphid farms. Birds flutter around the branches, eating the ants, and when the aphids are left exposed, predator insects come in and clean them up, helping the tree.

I went out to pick the raspberries tonight and a catbird landed not five feet away on a raspberry cane, and we both just froze. Catbirds are gray birds, not quite robin-sized, with a little black cap on their head. They're generally shy birds, mimicking other birds (they're cousins to the mockingbird) and sometimes making the *mew!* noises that give them their catbird name.

After a moment the catbird flew to the other side of the raspberry patch, picked one with its beak, tried to wolf it down whole but dropped it, and hopped to the ground to finish it off. Catbirds and robins like fruit. I don't mind sharing because I like having them around and they don't eat everything on the vine.

Now, squirrels on the other hand...! I've wrapped my apples in plastic bags in an attempt to keep some from the squirrels, which sometimes worked but often didn't. They're sneaky little jerks.

On the other hand, if you live in a place with a lot of wildlife, like an urban wildlife corridor, then you'll need to set up a plan for fencing in an area for yourself so you can have at least a little of the bounty!

The result should be a cute little garden (or a big one if you're ambitious) that you can hang out in, snacking on berries and making friends with birds and chipmunks whilst grousing at the damn squirrels who are eating all the hazelnuts. Hey, just because we are living with nature doesn't mean we can't cuss at squirrels.

Yeah, don't play innocent, bub, that's the 50ᵗʰ walnut you've swiped today.

PLANTING AND ESTABLISHING YOUR FOOD FOREST

Now that you've planned and designed your food forest, it's time to get your hands dirty and start planting! In this chapter, we'll discuss planting techniques, along with best practices for taking care of the food forest through the years.

The Logistics of Landscape Installation

So you like to get the work done. You've made your landscape plan, marked out the gardening beds in your yard, and scraped all the grasses from the planting area. You've moved all your outdoor furniture and other things out of the way, and you've installed your hardscape – things like a walkway made of pavers, a patio, a deck, or any permanent landscape features that your garden is going to fit around. If you decide to add an irrigation system, this would also be the time to have it professionally installed. (I'm skipping hardscape as this is outside the scope of my book.)

To prepare for your big planting day, get your compost and other organic soil amendments and spread them over the garden area, then till them into the soil. (If you prefer no-till gardening, skip this step and just dig large holes for each plant, adding the soil amendments into the soil around them.)

You've picked up all the plants, you've got the bags of mulch are standing by.

Plant trees and shrubs first, then fill in the gaps with smaller plants like herbs and groundcovers.

Group plants together in guilds to create a mutually beneficial environment.

Consider planting in phases, starting with the canopy layer and working your way down to the ground layer.

Preparing the Garden for Planting

As I've pointed out in other places in the book, I'm not a fan of using Round-Up or herbicides if I don't have to. I do have a quart of herbicide that I will use to spray poison ivy around the yard. But I use it carefully – do my best to spray only the poison ivy leaves and nothing else.

When it comes to clearing a garden space, I don't use the herbicide at all, and I don't pull up the weeds or till the soil.

I've found a neat trick that allows me to set up a garden area without herbicides, in a way that will knock down large weeds with very little effort while creating rich garden soil at the same time.

I mark the area where I want the garden to go. Then I'll take several large cardboard boxes, cut them along one side with a box knife to open them up, then lay them flat on top of the ground until the area where I want the garden is completely covered. I don't pull up the weeds, not even the big ones – I just throw the cardboard on top of them. If I need to get the cardboard to lie flat, I'll walk over it a few times until the big weeds give up.

If you don't have any cardboard boxes lying around, lay newspapers, 10 pages thick, over the garden area. You might have to put rocks or pile mulch on them in strategic places while you're laying them out, so they don't blow around in the wind.

Then I cover the cardboard with at least three inches of mulch (shredded leaves, wood chips, grass clippings, etc.). This smothers the weeds and give you an instantly clean landscape.

If your soil is especially poor, prepare the garden for next year in this way. Through fall and early winter, keep adding more mulch. Chop up fallen leaves with the mulching mower for plentiful, nutritious mulch. Then let it all break down through the winter.

Next spring, the soil should be ready for you to plant your food forest selections. When you move aside the mulch to plant, you'll be pleasantly surprised to see lots of rich soil under the mulch as well as a nice crop of worms to mix your new topsoil with the ground underneath.

Keep adding mulch, of course, all through the year. Feeding the soil isn't a one-time operation – it's a process. Mulch is the easiest way to keep up with that process.

However!

If your ground is especially dry, or poor, you will have to do more to prepare it – adding soil amendments and tilling them in – before you mulch.

Mark the area, then spread a thick layer of organic amendments such as coconut coir – an excellent, eco-friendly replacement for peat moss that does an amazing job at water retention – as well as compost or well-composted manure. You can also add some worm castings, blood meal, or bone meal, rock phosphate, kelp meal, or alfalfa meal to the mix, depending on what you have at hand, to add different minerals to the soil.

Spread all of this evenly across the area, then till or spade it in.

For compacted soil, or if you don't want all the racket that a tiller causes, you might try a tool called a broadfork. It's a good tool if you have hardpan that needs to break up – and if you have the muscles to use it and punch it into the heavy ground.

But in clay soil or rocky soil, a digging fork is a much better alternative, and you can use it one-handed! And it also can be used for other tasks, like pulling up carrots and loosening the soil around deep-rooted weeds to get them out.

Once the soil is loosened up and raked smooth into the contours you want, then throw down the cardboard and the thick layer of mulch as before.

Trees and shrubs need a little extra attention since you're planting them for the long term. Giving them a good start helps to give your plants get established and settle in nicely, even when the weather is against them.

Planting Techniques for Taproots and Lateral Roots

The problem with being a plant is that, if you end up in lousy soil or in a spot with bad drainage, you can't just haul out your roots and walk to a better location. They live the old adage, "Bloom where you're planted," because they have no choice! Digging a good hole is how you can help them get a good start.

Digging a proper hole takes time, especially if you don't handle a shovel very often, but when you place that time against the 30 to 50 to over 100 years that a tree could be growing there, it's time well spent.

Let's take a minute to look at how a tree grows. Contrary to popular belief, trees generally don't develop a gigantic taproot like a carrot. Now, most seedlings start with a taproot, because they want to anchor themselves in the soil as quickly as they can.

Young oak, hickory, walnut, and conifer trees will keep growing a big taproot when they're young. Ages ago, when I worked as a municipal horticulturist, we had to use a tree spade to move a number of burr oak and white oak trees. A tree spade is a machine with four large spades that close to a point around the roots of the tree, then you push a lever to bring the tree with its root ball out of the ground, held by the four large spades that were about four feet tall.

Sometimes we'd try to lift one of these oaks out of the ground, but they wouldn't come up, so we'd have to take the spades out, reposition the box around the tree, and send the spades down into the soil again, because the taproot on these young trees were so deep.

One tree we finally pulled out had about two feet of taproot sticking out from the bottom of the four-foot tree spades. Deano, who was helping me run the spade, was pretty impressed by this.

I was impressed by this too. When we later moved into our new home, it was at the end of a street where anybody with a lead foot and no eyes could drive straight into our yard. So I planted a white oak tree between the street and the house. We've had a few cars slide through the stop sign on snowy days and end up in the ditch. Nobody has gone farther than that – fine by me. And a white oak is a beautiful tree, so everybody wins.

At any rate, as a tree matures, it spends more energy on developing a lateral root system that spreads out through the top 12 inches of the soil. Here, the large roots stabilize the tree, while the tiny feeder roots soak up oxygen, nutrients from humus, and rain.

These lateral roots are not confined to the dripline of the tree – that is, the outer circumference of the tree's branches. Roots can spread a long way past this. This is why small trees should be planted at least three to four feet away from pavement (sidewalks, driveways, patios, etc.), and large trees at least eight feet away.

The Importance of Digging a Good Hole

So! With that in mind, let's get back to planting this tree.

First, dig a hole that's twice as wide as your tree's root ball, and just slightly deeper than the root ball, so the base of the root ball is sitting on solid earth.

Add some soil amendments to the hole, such as compost. Instead of adding peat moss, use coco coir, which is made from coconut hulls, making it a more sustainable choice. These amendments should be less than 20 percent of the backfill.

Once the hole is ready, set the tree in. Lay your shovel across the top of the hole, next to the trunk, so you can see if you've dug the hole deep enough. The rootball should sit under the shovel handle. However, if the soil consists of heavy clay or bedrock, plant the tree with the root ball slightly above the ground. If you're planting a grafted tree (many fruit trees are grafted onto hardy rootstock), then the bud union should sit above the shovel handle.

Remove twine and any exposed burlap (if you're planting a balled and burlapped tree) so water doesn't wick out of the ground.

Press the soil down around the tree, water it deeply to help settle the soil, and add in more soil when the ground sinks. Use the remainder of the soil to build a little dam around your filled-in hole to keep water in. Cover it all with two to four inches of mulch to protect the roots and soil.

MAINTENANCE FOR A THRIVING FOOD FOREST

Essential Watering Strategies

While a food forest is designed to be a low-maintenance system, the act of watering can be a high-maintenance act in the early years as you get the plants established. However, there are ways to streamline the process.

In the early years, newly planted trees and plants will require regular watering, particularly during dry spells.

Water deeply and infrequently to encourage deep root growth. For example, with my newly-planted apple tree, and my older established trees, I just stand there with my hose running straight into the mulch for a while, and just soak the heck out of it, especially when it's dry.

Deep and infrequent waterings, in theory, allow the water to soak deep into the ground, and then the roots of your trees and plants reach more deeply into the ground. Roots too close to the surface of the ground get hot and dry out when the ground dries out in the heat of the sun (though a nice layer of mulch will also keep the soil cool and moist).

On the other hand, if you have a big garden, that's a lot of time to spend walking around with a garden hose. Streamline the work. Invest in a soaker hose or drip irrigation system to minimize water waste and ensure even watering.

Drip tape looks like a flattened hose with small holes for the water. Drip tape works even when your water is coming out at very low pressure (so you can just turn your water on only a little bit and still get your crops watered).

Put the drip lines or soaker hoses next to the plants, then cover them with mulch to keep the water cool and minimize evaporation.

Maintaining Soil Health and Fertility

The health of your soil is critical to the success of your food forest. With a focus on perennials, your soil will benefit from the ongoing presence of living plants, which can help to build soil structure, increase organic matter, and promote microbial activity. Consider planting nitrogen-fixing species like clover, lupine, and beans, which can help to improve soil fertility naturally. In addition, top-dress with compost and organic matter annually to ensure your soil remains healthy and productive.

However, the best and easiest way to good soil is through a nice, thick layer of organic mulch.

Mulching is insanely important for maintaining a healthy and productive food forest. The best mulches are organic – able to break down and nourish the life in the soil.

I'm personally a huge fan of mulch, and in every book I've said the same things about mulch. So! instead of going over it again, I'm going to have Vince McMahon of the WWE explain the benefits of mulching.

Adds organic material!

Prevents erosion and splashback!

Protects roots from hot sun!

Holds moisture in the soil!

IMPROVES THE WORLD!!

Thanks for the update, Vince.

Another important factor to consider is soil pH. Most plants prefer a slightly acidic soil pH, between 6.0 and 7.0. If your soil is too acidic or alkaline, you can adjust the pH by adding lime (to raise the pH and increase alkalinity) or sulfur (to lower the pH and increase acidity).

In addition to improving soil quality, it's important to provide your plants with regular fertilization. Organic fertilizers, such as compost tea or fish emulsion, are a great choice, as they provide slow-release nutrients that won't burn your plants.

Add organic matter, such as compost or well-rotted manure, to improve soil structure and nutrient content.

What about biochar?

Biochar is an old fertilization method that's come back into vogue. It was used by the folks in the Amazon for ages as a soil amendment, along with bones, compost, and manure. It's an amazing addition for tropical soils. Biochar is basically wood charcoal created with low-intensity smoldering fires. Before the Spanish invaders showed up and everything went to hell, the folks in South American added biochar to an area that was twice the size of Britain, creating a rich black soil that still holds a great deal of carbon to this day, and provides home for microbe and fungi, which increases soil biomass (also a win for plants). It holds water and nutrients very well.

However, in temperate climates, biochar doesn't seem to have the same effect on the soils. It works if you have sandy or sticky clay soils, or soil low in organic matter. It's a good addition to greenhouse potting media. However, the pollution created by the charcoal creation process is a substantial drawback.

Biochar is still great at holding water and increasing the cation exchange capacity of your soil (that is, the soil's ability to hold and supply nutrients in the form of positively-charged ions).

On the other hand, some plants are negatively affected by biochar, because it will raise the soil pH, so plants that enjoy acidic soils (conifers, blueberry plants, magnolias, etc.) get stressed in biochar-amended soils. Some microbes and fungi are also sensitive to pH changes that biochar causes.

When in doubt, take a soil test and talk to your local University Extension agent or a soil scientist who is familiar with local soils.

Pruning and Thinning Techniques

Pruning is an essential part of caring for fruit trees and plants in a food forest. By removing dead or diseased wood, shaping the tree to encourage good airflow and light penetration, and managing the overall size of the tree, you can help to ensure healthy growth and bountiful harvests.

Prune your trees and shrubs during the dormant season to minimize stress and promote vigorous growth.

Thin out weak or diseased branches to prevent the spread of pests and diseases.

Pest and Disease Management

Pest and disease management is an ongoing task in any garden, and the food forest is no exception. Here are some tips.

1   Use integrated pest management techniques such as companion planting and biological controls to minimize the use of pesticides.

2   Monitor your plants regularly for signs of pests and diseases and take action early to prevent any problems from spreading.

3   My main approach to pest control is keeping an eye on my plants and knocking back pests by blasting them with a water hose, and squishing them, or stomping on them. It helps to recognize when and where pest eggs appear, and what they look like, so you can nip an issue in the bud.

4   I avoid pesticides that could kill pollinators because I want to create a diverse and healthy ecosystem so predatory insects and birds do this work for me. When Japanese beetles show up, I knock them into a cup of water, then give them to my chickens. You can reduce the risk of pest outbreaks, while also using natural interventions like beneficial insects, trap crops, and hand-picking to manage any issues that arise.

5   To be honest, once you've stopped using herbicides and pesticides, you'll get a lot more help in the garden. Leave it to us civilized humans to be so insanely controlling as to claim a smidgeon of land as exclusively "ours," and killing any insect or plant that we deem unworthy of being here because this is *our property*. It's just wild.

Succession Plantings for Continuous Harvests

Succession planting involves staggering the planting of different crops at different times, ensuring a continuous harvest throughout the growing season. This approach can be particularly useful in a food forest, where different trees and plants will mature at different rates.

For example, you might plant early-season crops like radishes and peas beneath young fruit trees, then follow with mid-season crops like squash and beans, and finish with late-season crops like kale and carrots.

You can also extend the harvest window by seeding the same crops over successive weeks – like seeding lettuce every 7 or 14 days so you have a continual harvest, instead of having the lettuce all grow up at the same time and now you're done.

So that's succession plantings.

Your challenge is to find ways to add different food plants so the harvests keep rolling along.

I recommend keeping a gardening journal, and writing down the dates when the different crops are planted, when they bloom, and when they're ready to harvest. Do this for all your plants. When you do this for even just one year, you start seeing gaps in harvest that you can fill.

So this year, you get the mulberries and strawberries, followed by the black raspberries. Then maybe there's a little gap before the tomatoes come in – but that's because you started them late, perhaps, or perhaps it's because the place they were planted is getting too much shade.

So next year you're like "Okay, I've got to get the tomatoes started earlier, and this time, let's plant them over by the shed so they can get more sun." You mark all this on next year's calendar so you can remember, and then boom, you are officially ahead of the game.

This is also how your garden grows and gets better through the years.

Pollination

Fruit trees and plants rely on pollinators to set fruit, so caring for the bees and other pollinators is an important part of food forest care.

While many native bees and other pollinators will naturally visit your food forest, you can also encourage them by planting flowers that provide nectar and pollen throughout the growing season.

Consider adding native wildflowers like asters, goldenrod, and milkweed, which are good for pollinators but are also food crops for caterpillars – which turn into butterflies later on.

PLANT PROFILES FOR YOUR FOOD FOREST

Fruit Trees

Notes on fruit trees in general:

1) If you notice your branches are too heavily-laden with fruit in late summer, thin out the small and insect-damaged ones. The remaining fruits then can grow larger and this keeps your overburdened branches from breaking. Or give your branches support by either staking them, using a piece of lumber with a V cut into one end, or tying it to an overhead branch.

2) To improve winter pruning, take some pictures of all four sides of your fruit tree in fall. Make notes of spaces that need more air circulation, limbs that aren't bearing, and so forth. When you get ready to prune your tree in winter, refer to these pictures and use this time to fix what needs fixing.

3) Be sure to clean up fallen fruit and leaves under the tree to keep diseases and pest insects from overwintering. Also, cleaning up fallen fruit keeps yellow jackets out of the yard. I have no problem with honeybees enjoying rotten fruit. But when it comes to yellow jackets and hornets, I'll see them all in hell!

Apples

Apple trees are a popular and productive addition to any food forest setup. They can provide delicious fruit and beautiful spring blossoms while serving as an important component of the ecosystem.

Here's an overview of how to grow apple trees in a food forest:

Height and Width: Apple trees can grow to a height of 20 to 30 feet and a width of 15 to 25 feet.

Cultural Requirements: Apple trees require full sun, good air circulation, and well-draining soil with a pH between 6.0 and 7.0. They also need regular watering and fertilization.

Pests and Diseases: Apple trees can be susceptible to a number of pests and diseases, including aphids, apple maggot, codling moth, fire blight, and powdery mildew.

Planting: When planting apple trees in a food forest, choose a spot that receives full sun and good air circulation. Plant the trees in a hole that is wider than the root ball and deep enough to cover the roots. Mix compost into the soil to improve drainage and fertility.

Mulching: Apply a layer of mulch around the base of the tree to conserve moisture and suppress weeds. Choose a mulch that will break down over time and add organic matter to the soil.

Fertilizing: Apple trees benefit from regular fertilization. Use a balanced fertilizer in the spring and mid-summer to encourage healthy growth and fruit production.

Pruning: Prune apple trees in the late winter or early spring before new growth appears. Remove any dead, diseased, or damaged wood, as well as any branches that cross or rub against each other. Prune to shape the tree and promote good air circulation.

Harvesting: Apples are ready to harvest when they are fully colored and easily twist off the tree. Store them in a cool, dry place to extend their shelf life.

Pretty apple blossoms in spring.

Mulberries

These are considered a weed tree in many places, but if you have a yard where it's hard to grow plants, this fast-growing tree will likely prosper. But give it space. A mature tree can grow 30 to 50 feet tall, sometimes up to 70 feet, and the canopy is about 40 feet wide. Makes a great shade tree, but don't park your chair directly under it during fruiting season due to all the purple bird poop.

If you're in an area with not many mulberries around, you'll need a male and a female tree – so you'll have one tree that grows only pollen and a second tree that grows the flowers and the fruit.

However, if you know how to graft, you can graft branches from a male tree onto a female tree so you can get the benefit of pollination while growing only one tree.

I eat mulberries straight off the tree. Lay an old bed sheet or tarp under the tree and shake the tree or let them fall naturally through the day. Chickens also like them. I have a small, white mulberry tree growing in the chicken pen for the girls. I keep it trimmed back hard so it doesn't get any taller than six feet and it stays out of our way. It provides shade and it's already providing berries.

Great for pie, jams, jellies, preserves, compotes, sauces, wine, dehydrating, soda. Put them in muffins, pies, pancakes, quick bread. Goes great in a rhubarb pie. You don't have to remove the little stems as they're edible, but they might be an annoyance in the pie.

The juice will stain you and your clothes – but if you make your own dye for yarn or cloth, mulberries are the bomb. But you also have purple poop from birds that feast on the berries. It's hell if you hang your clothes on a clothesline to dry during mulberry season. There's always a bird that will perch on the clotheslines and leave a purple poop trail down your nice white shirt.

You can use the wood to smoke meat, and Native tribes used the inner wood fibers to weave soft fabric.

The red mulberry (*Morus rubra*) is native to North American, while the other mulberries are native to different parts of the world.

Pawpaws

Pawpaw trees (*Asimina triloba*) are native to the eastern United States and produce unique and delicious fruit with a tropical flavor. Here is a guide on how to grow pawpaw trees in a food forest setup:

Size: Pawpaw trees can reach a height of 20 to 30 feet (6-9 meters) and a width of 15 to 25 feet (4.5-7.6 meters) at maturity.

Cultural Requirements: Pawpaw trees thrive in well-drained soils that are rich in organic matter. They prefer slightly acidic soil with a pH of 5.5 to 7.0. They require regular watering during the growing season, especially in the first few years after planting. Pawpaw trees also require some shade during their early years but will grow best in full sun once established.

Pests and Diseases: The biggest pests of pawpaw trees are the zebra swallowtail butterfly and the pawpaw sphinx moth, which lay their eggs on the leaves of the tree. While these insects can defoliate the tree, it is usually not a major problem. Diseases that can affect pawpaw trees include Phytophthora root rot, which can be avoided by planting in well-drained soil, and black spot, which can be treated with fungicides.

Growing Pawpaw Trees: Pawpaw trees are typically grown from seeds, which can be purchased from nurseries or harvested from mature fruits. Seeds should be planted in a well-drained soil mixture and kept moist until they germinate. Once the seedlings are large enough, they can be transplanted to a permanent location in the food forest. Pawpaw trees require minimal pruning, but dead or damaged branches should be removed.

Taking Care of Pawpaw Trees: As mentioned earlier, pawpaw trees require regular watering, especially during the first few years after planting. It is also important to maintain a layer of mulch around the base of the tree to retain moisture and suppress weeds. Fertilization is usually not necessary, as pawpaw trees are adapted to low-nutrient soils. However, adding compost or a balanced fertilizer can help promote healthy growth. In addition, pawpaw trees can benefit from the presence of nitrogen-fixing plants, such as legumes, in the food forest.

Nut Trees

Include some nut-producing species to provide protein. You can get small-stature or slow-growing species, such as hazelnuts, filberts, nut pines, or get crazy with big trees such as pecan, walnut, or hickories.

Chinquapins and Chestnuts

While researching this book I was surprised to run across two dwarf chestnut species that are apparently alive and well.

The Allegheny chinquapin (*Castanea pumila*) can be grown in zones 3-9 as a shrub or a dwarf tree that gets up to 12 to 15 feet tall. The Georgiana chestnut (*Castanea alnifolia*) only gets up to 4 feet tall in zones 8-10 – very much a Southern plant.

I thought, well, these are likely regular chestnut trees that would grow to full height if chestnut blight were not still around. However, chestnut trees start growing, and then blight steals through the neighborhood and kills off the young tree, which then regenerates from the roots. This is what happens to most of the American chestnut trees.

But it turns out I am wrong about the dwarf chestnuts, apparently! These are actual dwarf trees that bear sweet chestnuts around October, though you have to fight off deer and turkeys and rabid squirrels and sometimes your neighbors for the nuts.

The American chestnut used to grow as a gigantic tree up and down the Appalachians and through a broad swath of the United States, and the nuts fed a lot of people. They made a tasty snack, and they could also be ground into flour and used in a million different ways. These trees often grew to be over 100 feet tall and were more prolific with producing nuts than oak trees, which is pretty wild.

At about the turn of the century, (that is, about 1901) imported chestnuts brought along a fungal disease – chestnut blight – which spread like wildfire through the chestnut population, killing between 3 and 4 *billion* trees. So that's the American chestnut.

However, the Allegheny chestnut is a true dwarf species that is more resistant to the blight (though it will succumb now and then). Often Alleghenys affected by chestnut blight will stubbornly keep putting out suckers to produce fruit. Scientists and breeders keep working to identify and cross resistant varieties to increase blight resistance.

Side note: The Ozark Chinquapin Foundation is doing exactly this for the extremely rare Ozark chinquapin, *Castanea ozarkensis*. If you sign up for an annual seed membership, they'll send you a little bunch of chinquapin seeds in January to plant so you can grow your own Ozark chinquapin. Each year they improve blight resistance in these trees a little bit more. A worthy cause indeed! at https://ozarkchinquapinmembership.org.

Shrubs

Berries such as raspberries, blackberries, blueberries, and strawberries can be grown in beds or on trellises. Consider adding more exotic fruits like kiwi, persimmons, and figs to your food forest for a unique touch.

Brambles

Raspberries, my absolute favorite

Bramble fruits, such as raspberries and blackberries, are a great addition to any food forest. These hardy plants can thrive in a variety of soil types and are known for their tasty and nutritious fruits.

Height and width: Bramble fruits vary in size depending on the specific variety, but most can grow up to 6 feet tall and spread out several feet wide.

Cultural requirements: Bramble fruits prefer well-drained soil and full sun but can also tolerate partial shade. They benefit from regular pruning to remove old canes and promote new growth. Brambles can be propagated by tip layering, root cuttings, and stem cuttings.

Pests and diseases: Bramble fruits are susceptible to several pests and diseases, including aphids, spider mites, powdery mildew, and raspberry crown borer. Proper pruning and care can help prevent these issues.

Growing bramble fruits in a food forest is relatively easy. Plant them along the edges of the forest or in a sunny spot with well-draining soil. Provide support for the plants with trellises or stakes, as they can become top-heavy when bearing fruit. Fertilize with compost or organic fertilizer to encourage healthy growth and fruit production.

When harvesting bramble fruits, it is important to do so carefully as the plants have thorns. It's important to create some kind of path through them so you can step into them and get the fruit way off in the back without getting scratched all over. Bramble fruits can be eaten fresh or used in jams, jellies, and baked goods. Harvesting regularly can help encourage new growth and a longer fruiting season.

Ribes species are good at fruiting in the shade – heck, they'll root themselves in a shady area if you'll let them.

Roses

I might be something of a rose fiend, but I can't help it. Roses have such a long history, stretching back to the time of the Pharaohs and even before, and they smell glorious.

Roses can also work well as a wild hedge, as a place for wildlife to live in safety (if you don't mind a few bunnies and chipmunks). And they're a fragrance dispenser on hot days when the roses are blooming like crazy.

Roses, like the heirloom 'Autumn Damask' above, are edible in every part! Young rose leaves and rose buds can be used, fresh or dried, to make an herbal tea that tastes something like black tea (except without the caffeine).

Rose petals are used in cuisines all over the world in an amazing array of different uses. Add them to a salad. Or chop the petals finely, mix them with a little honey, and use it as a spread.

Rose hips, once they turn orange or red, can be stewed like apples. Cut them in half, scrape out the seeds, and use them for jams and jellies, sauces, or fruit leather.

If you want a rose to eat, choose the old-fashioned, strongly fragrant varieties. I have a huge list in my _Rose to the Occasion_ book, which is all about roses, but here are a few of my favorites. These roses are all easy to grow, hardy as heck, and have a delicious fragrance.

I grew all of these in the municipal rose garden when I was singlehandedly running it, and they were so easy to care for! They made my life as a horticulturist a little less stressful.

Blanc Double de Coubert
Thérèse Bugnet
Hansa
Carefree Delight
Autumn Damask (aka Quatre Saisons)

Perennials

Perennials and plants that reseed themselves (and are true to seed) help to reduce work. Perennials also are more effective at capturing nutrients that leach out of the soil than annuals are. And you don't have to plant perennials every year – they stay put and stay alive through the winter. I like perennials because they always save me a lot of labor in the spring!

Perennial Roots and Tubers

Here is a scattering of root vegetables that taste good and grow in semi-wild, semi-shaded areas with a minimum of fuss.

Root crops like sweet potatoes, yams, and Jerusalem artichokes can be grown as perennials in a food forest. These crops require minimal maintenance and can provide a bountiful harvest.

Skirret was a popular root plant, clear back to the days of the Romans. Cooked, the roots taste like a mix of carrot and potato, sweet and soft, but take out the woody center core as you're eating so you don't end up with little woody threads in your teeth. The rest of the root is delicious. The spring shoots are also tasty. Grow this as a perennial in shade or forest edges; this plant will fend for itself.

Salsify was a favorite tuber of Victorian times. It features a slender parsnip-like root that can be roasted, mashed, or boiled. It's a member of the lettuce family, and its early spring shoots are edible. Harvest roots in late fall to early winter – but dig them up before the ground freezes. Boil the roots before peeling them, or peel them in a bowl of water, because otherwise the roots will gum you up with a lot of sticky latex. (This is why the plant is called kitchen maid's disaster in the Netherlands!) Once peeled, it's wonderfully versatile and tastes like artichoke hearts.

Even though salsify is technically a biennial, you might be able to grow it in a perennial bed because it can regenerate from the roots snapped off during harvest.

Scorzonera is grown and prepared much as salsify is. This plant has crazy black roots that reach up to three feet in length.

Cosnes – *Stachys affinis* or Chinese artichoke – is a funny little tuber that looks like a crinkle fry or a large piece of pasta. They grow as a mat of vegetation that's 12 to 18 inches tall, and can get invasive, but the taste of these little tubers is wonderful. Chefs love their versatility, but they're hard to get (and therefore expensive) due to their short shelf life. Keep them watered to encourage the tubers.

If voles and moles are a problem, raise them in big planting boxes with hardware cloth laid along the bottom to keep hungry critters out of it. Harvest after the foliage dies back in November.

Incorporating a diverse range of perennial crops into your food forest can provide a steady source of food throughout the year. By selecting a variety of crops, you can ensure a balanced and nutritious diet. Don't be afraid to try new crops and experiment with different planting techniques to create a food forest that suits your needs and preferences.

Annuals

Integration of Annual Crops

While the focus of a food forest is on perennial crops, annual crops can also be integrated into the system. Consider planting crops like lettuce, peas, and beans in between perennial crops to make use of available space and provide variety in your diet.

Really, anything that self-sows once it goes to seed has a viable place in the food forest.

Actually, not *anything*. Plants that are invasive should not be given a place in your garden. Choose plants that will spread cheerfully but not aggressively.

Carrots

Carrots will self-sow, but after about three generations they'll revert to their wild "roots," as it were. So be sure to buy new carrot seeds every few years. They'll fit into all kinds of guilds, and the same goes for radishes. But most of these other root crops grow semi-wild and persist over long periods through self-sowing.

Radishes

Also a very handy and easily-seeded crop that can make itself at home in many situations. Let it bolt and go to seed.

Herbs and Medicinal Plants

Herbs like thyme, sage, mint and add a savory element to any dish. Medicinal plants like echinacea, ginseng, and comfrey can be grown for their healing properties. Comfrey is planted all over food forests. Herbs and medicinals can be added to random places for scent and utility.

Wild Greens

In early spring, and sometimes in late spring, Grandma Mary would put on plastic kitchen gloves and cut full-grown nettles to eat. I generally stayed far away from this operation, having met many nettle plants in my life that would inflict itchy, burning red dots on my bare legs or arms when I didn't see them quickly enough. But she'd cook them down in a cast-iron skillet with a little bacon grease. Apparently, these made good greens, though younger me passed them by on Sunday dinners.

Greens, whether wild or domesticated, make a great groundcover (well, not the damn nettles) and can be picked and eaten if you forgot to pick up the lettuce at the store. Hell, who needs lettuce if you have Swiss chard brightening up the sunny forest edges?

Mushrooms

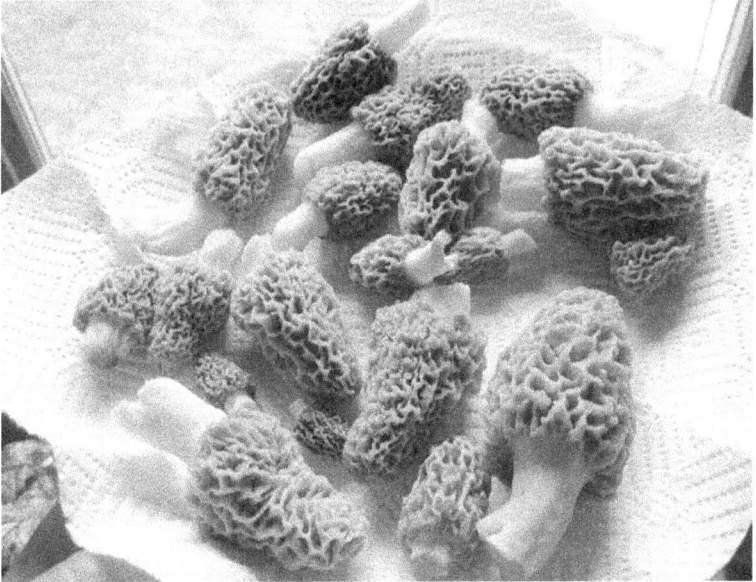

Morels!!!!!!!!!!!!!!

"About 65 times more energy goes to decomposers than goes to all herbivores combined, except when leaf-eating insect populations explode" (not literally). *Edible Forest Gardens, Vol. 1., p. 148*

There's an easy, and very tasty, way to capture all that decomposer energy and not let it go to waste: By planting mushrooms!

King stropharia, shiitake, oyster, chicken-of-the-woods, reishi, wood ear, lion's mane, and shaggy mane mushrooms are all popular varieties that foragers love to gather in the woods.

And of course morels, the King of Mushrooms, pulls Midwesterners of all stripes and ages into the local forests in April and May. Facebook feeds sprouts morels as everybody shows off their day's haul from the woods (and occasionally from some shady spot in their backyard).

Mushroom gardening is relatively unexplored territory for most gardeners, but you don't need to be a mycologist to grow your own.

To grow mushrooms, you need spawn, which contains the mushroom mycelium that is waiting to be activated to grow. Spawn can be bought in a bewildering array of forms – myceliated grain, or sawdust, or little wooden plugs and pegs. These contain various forms of spawn, and you place them placed into the places where they will grow.

Here are a few of the more common mushroom spawns.

- **Plugs** are wooden dowels colonized with mycelium. You drill a series of holes into the log where you intend to grow the mushrooms, then tap the plugs into those holes. They're sealed in with a little wax to keep them from drying out.
- **Grain spawn** is used to inoculate straw, but this needs to be used in a place where it doesn't attract birds or rodents.
- **Sawdust spawn** will carry fungi that naturally grow in logs, and can sprout mushrooms more quickly, but must be watched to be sure it doesn't dry out.

Mushroom spawn can grow in beds of wood chips or straw mulch in your forest. Full shade is fine – anyplace where plants can't grow, where the substrate won't dry out, is perfect for mushrooms.

So, for example, you can put down a light layer of wood chips, sprinkle the spawn over them, then cover with another layer of wood chips, and water them in. And lo! You have created a mushroom bed.

Plugs can be tapped into new logs or freshly-cut tree stumps, allowing the fungi to take over the wood and grow.

Important note: Not all logs or trees are good for mushroom growing! Use logs that have been cut and aged for at least several weeks – long enough for the cells of the tree to die off, but also to still have moisture in the log for the fungi to live off of.

Start planting your mushroom varieties once the daytime temperature consistently stays over 40 degrees F so you can get mushrooms before frost.

Mushrooms are a great way to use those areas of dense shade where nothing else will grow.

FREEZING AND DRYING

Proper harvesting and storage of your food forest crops is important for preserving their quality and flavor. Some crops, like berries and fruits, can be preserved by freezing, drying, or canning for later use.

Miss Sophie shows us how to dry strawberries.

I haven't done much on the canning side of things, but I invested in a dehydrator a couple of years ago. I use it almost exclusively for drying apples, but it's great for other surplus fruits and vegetables.

Store your dried fruit in a glass Mason jar or a freezer Ziplock bag with the date and variety written on it. I keep the dried fruit in the freezer, tightly sealed, for even longer shelf life. Also, if I leave the bags in the cabinets, I tend to eat through the dried apples pretty quickly. Oops.

Yum!

My Aunt Candy will cut the apples up just as I do, but she freezes them as apple pie filling. She uses her favorite apple pie recipe to mix them with cinnamon and sugar and other good things, puts enough apples in a freezer bag for a pie, and pops them in the freezer until she needs them.

Raspberries and other berries can be frozen. After washing and draining the berries, I spread them out briefly on paper towels to get the excess water off them, then spread them in a single layer on a piece of aluminum foil on a baking tin. I pop them into the freezer for 12 to 24 hours to freeze solid, then pour them into a gallon-sized freezer bag. They'll keep for over a year this way, and when you shake out a couple of berries they sound like marbles rattling.

*

Thus endeth the book on Food Forests!

If you liked this book, do give me a review, and look at the next book, wherein we make a mad dash into our wild areas to find our own awesome food.

BOOK 6 OF THE HUNGRY GARDEN SERIES

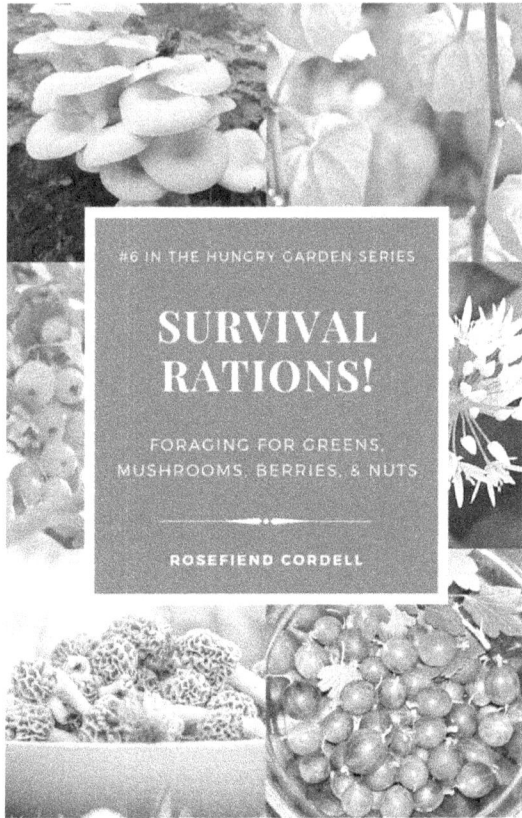

Preorder your copy here!
https://books2read.com/u/mV0wzp
I'm not sure when I'm releasing this, but I'll try to keep
you-all posted.

OTHER BOOKS IN THE HUNGRY GARDEN SERIES

Little Pots, Big Yields: Container Gardening for Creative Gardeners

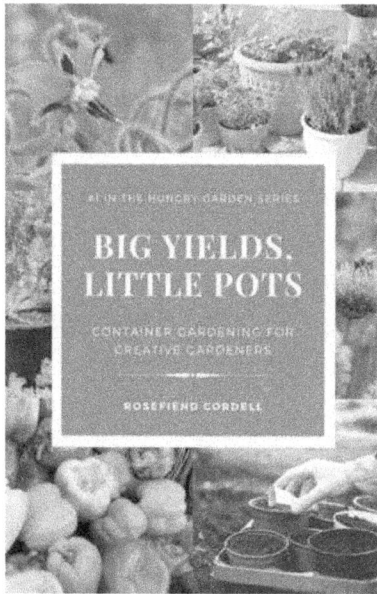

🌱 Discover the Secrets of Bountiful Harvests in Little Pots! 🌿

Are you craving a thriving vegetable garden but limited by space, challenging soil conditions, or physical constraints? Fear not, for "Little Pots, Big Yields: Container Gardening for Creative Gardeners" will unlock a world of fresh herbs, succulent tomatoes, and luscious strawberries right at your fingertips.

Think Big in Small Spaces: Let the magic of container gardening revolutionize your green thumb journey. Maximize your limited space and transform balconies, patios, and even windowsills into havens of homegrown goodness.

The Art of Container Gardening Unveiled: Delve into the essential aspects of container gardening, from selecting the perfect containers to combating damping-off disease. Master the secrets of soilless mixes and the crucial elements for plant success. Discover the fine balance of fertilizer, watering, climate control, and the art of trellising.

From Seed to Savory Delights: Dive into a cornucopia of vegetables that thrive in pots and unlock the best methods for each crop, including the best varieties that will flourish in your little pots. Whether you're a beginner or a seasoned gardener, this indispensable guide will empower you to yield bountiful harvests.

Embrace the Green Revolution: Little Pots, Big Yields is your ultimate companion, whether you're a container gardening novice or a seasoned pro. Discover the secrets, tips, and tricks that will propel your garden to new heights of productivity and creativity.

Are you ready to witness the magic of growing your own vegetables, even in the tightest of spaces? "Little Pots, Big Yields: Container Gardening for Creative Gardeners" will let your little pots flourish and reap the big rewards of a bountiful container garden!

Edible Landscaping: Foodscaping and Permaculture for Urban Gardeners

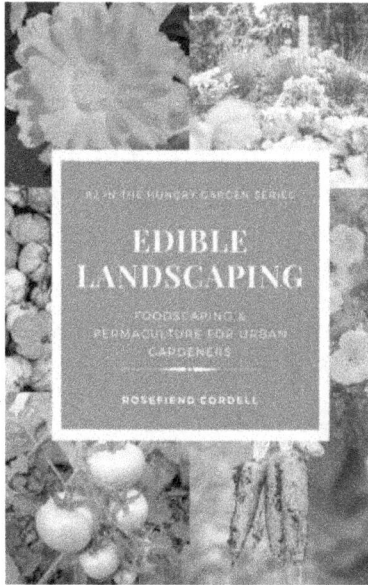

🌿 Unlock the Art of Edible Landscaping: Transform Your Yard into a Delicious Paradise! 🌱

Say goodbye to mundane vegetable gardens and say hello to a vibrant oasis that tantalizes your senses! In "Edible Landscaping: Foodscaping and Permaculture for Urban Gardeners," Rosefiend Cordell presents a revolutionary guide to turn your lackluster yard into a breathtaking sanctuary where beauty and bounty intertwine.

🌸 Feast for the Eyes and Palate: Immerse yourself in the step-by-step process of creating an enchanting garden where ornamental flowers elegantly dance alongside lush tomatoes, fragrant herbs, and tantalizing edible flowers. Unleash a cornucopia of flavors and elevate your living space into a symphony of colors, textures, and tastes.

🌳 Cultivate an Outdoor Haven: Discover the secrets of foodscaping and permaculture as you embark on a journey to build the garden of your dreams. Learn to nurture the soil, master garden design, and carefully select the plants that will thrive in your unique space.

🌿 A Harmonious Melody: Embrace the magic of a mixed border, where herbs, edible flowers, vegetables, ornamental plants, and fruitful trees coexist in perfect harmony. Whether you possess a green thumb or consider yourself a novice, this indispensable manual empowers you to unlock the potential of your garden.

🌾 Blossom Where You're Planted: Irrespective of your gardening background, this handy-dandy guide will unveil the secrets of maximizing your available space. Unleash your creativity and embark on a timeless gardening journey that not only captivates the eye but delights the palate, offering a deeply satisfying and rewarding experience.

Are you ready to revolutionize your gardening approach? "Edible Landscaping: Foodscaping and Permaculture for Urban Gardeners" is your passport to a world where artistry and sustainability meet. Unleash the potential of your outdoor space and savor the incredible delights that await you! 🌸

Beneficial and Pest Insects – The Good, the Bad, and the Hungry

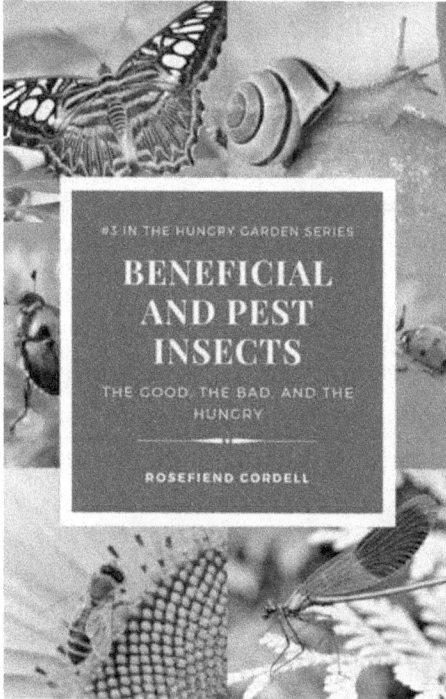

Uncover the Secret World of Garden Guardians and Pests! 🐞

Step into the mesmerizing world of insects thriving within your very own garden! *Beneficial and Pest Insects: The Good, the Bad, and the Hungry* delves deep into the captivating lives of these tiny creatures, from the plant protectors to the mischievous marauders.

Learn the Art of Observation: Discover the fascinating tales of the insects that call your roses home and the secretive egg-layers beneath your tomato leaves. Explore the hidden world of ambush predators, lying in wait to defend your precious plants.

Allies and Adversaries Unveiled: Master the art of identifying your allies from your adversaries with expert guidance. Embrace the beneficial insects like damselflies, ladybugs, praying mantises, and wasps while tackling the notorious mealybugs, aphids, stink bugs, and more from the rogue's gallery.

A Symphony of Nature's Harmony: Unleash nature's balance by encouraging the beneficial insects and discouraging the pests, using safe and chemical-free techniques. Become a virtuoso in orchestrating a garden symphony where your plant protectors reign supreme.

Home Sweet Habitat: Unlock the secrets to creating a haven for your insect friends. Attract the right crowd by providing them with the perfect habitat. From cozy lodgings to delectable dining, you'll have them buzzing with happiness in no time!

Cultivate Your Green Sanctuary: Discover sustainable methods to control pests without resorting to harmful chemicals. Embrace eco-friendly solutions that nurture both your garden and the environment.

Don't miss out on the enchanting world of insects right in your own backyard! Beneficial and Pest Insects is a must-have guide for gardeners of all levels. Get ready to witness the dance of nature's smallest warriors and transform your garden into a thriving sanctuary of life!

Indoor Gardening: Growing Herbs, Greens, & Vegetables Under Lights

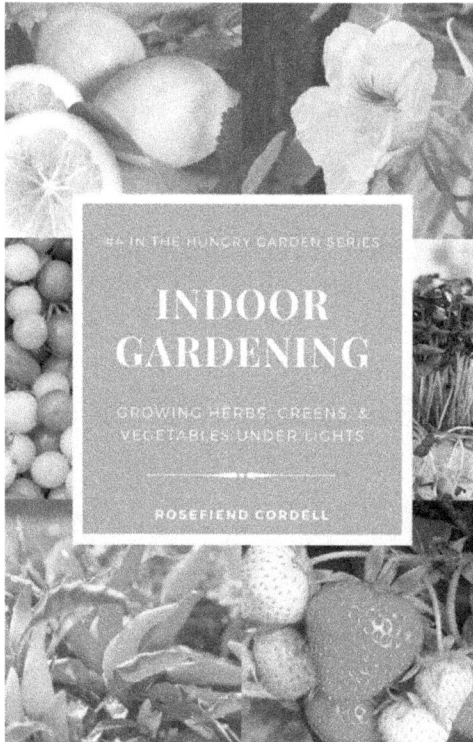

🌿 Discover the Thriving World of Indoor Gardening: Cultivate Fresh Delights Under Lights! 🌱

Gardening enthusiasts, rejoice! "Indoor Gardening: Growing Herbs, Greens, & Vegetables Under Lights" unveils the secrets to growing an abundance of organic vegetables, fruits, and greens right in the comfort of your own home—using affordable materials you already have!

🏡 From Kitchen Windows to Bookshelf Gardens: Embrace the joy of cultivating your own fresh, healthy produce without the need for land. With just a few well-placed grow lights and a dash of creativity, you can turn any corner into a vibrant haven of delectable delights.

Sprouts, Microgreens, and Herbs Galore: Delight in the art of indoor gardening as you learn to cultivate sprouts, microgreens, and a cornucopia of herbs that will liven up every meal. Imagine having your own little salad bar or an array of fresh herbs to add a burst of flavor to your dishes — all from the comfort of your home.

Uncover the Uncommon Indoor Crops: Dive into a captivating world where peas, tomatoes, and strawberries thrive indoors. Unravel the secrets of hand-pollination to ensure the success of your indoor crops.

Guided by a Horticultural Guru: Rosefiend Cordell, a horticulturist turned gardening writer, leads you through the exciting journey of indoor gardening. Learn the art of making a perfect space for your plants, selecting the right containers and potting mix, and understanding the crucial role of grow lights.

Overcome Challenges: Arm yourself with knowledge on how to tackle common pests and diseases that may try to invade your indoor oasis. Be empowered to nurture your plants to perfection with expert guidance.

Are you ready to transform your living space into a thriving garden of fresh, organic goodness? "Indoor Gardening: Growing Herbs, Greens, & Vegetables Under Lights" is the ultimate guide to cultivate a flourishing indoor paradise.

Me with a trowel and cosmos at Grandma Ann's, probably 1976

ABOUT THE AUTHOR

A former city horticulturist and a long-time garden writer, Rosefiend Cordell, aka Melinda R. Cordell, has written 12 books in the Easy-Growing Gardening series under the name Rosefiend Cordell, and three books (so far) in the Hungry Garden series.

She's worked in horticulture for half of her life – longer if you count when she was young, collecting wildflowers. She's worked in greenhouses, both retail and commercial; as a landscape laborer and designer; as a perennials manager; as municipal horticulturist and public rose garden potentate; and now as a gardening author (which is much easier on the back and joints).

Melinda R. Cordell has written a truckload of YA novels, including the Dragonriders of Fiorenza series. Set in an alternative medieval Italy, it features a wily dragonrider, her loyal dragon, and her assassin grandma, all pitted against a world out to strip away every one of their hopes and dreams.

Melinda lives in northwest Missouri with her husband and two kids, the best little family to walk the earth, and is writing about 24 books at once, fueled by passion and caffeine.

If you want to keep up with her, you can drop her a friendly note at rosefiend@gmail.com.

Don't forget to leave a book review on your favorite retailer, BookBub, or Goodreads!

melindacordell.com

www.ingramcontent.com/pod-product-compliance
Lightning Source LLC
Chambersburg PA
CBHW032003080426
42735CB00007B/497